EXTRACTION!

Comix Reportage

Edited by David Widgington, Frédéric Dubois & Marc Tessier.
Book design by Marc Tessier & David Widgington.
Front cover illustration and page 8 by Alain Reno.
Inside cover illustration and spot illustrations by Carlos Santos.
Back cover and chapter illustrations by Jeff Lemire.
Photographs by Marc Tessier except where indicated on page 5 & 115 by Dawn Paley and on page 115 by Tamara Herman and Brook Thorndycraft. Original version of *From the bottom of the pit* was published in *Canadian Dimension Magazine* (Vol. 40, No. 4, July/Aug. 2006, pp26-36).

Legal Deposit, Library and Archives Canada, 4th quarter 2015
Library and Archives Canada Cataloguing in Publication

"Extraction! Comix Reportage" edited by David Widgington, Frédéric Dubois, Marc Tessier.

ISBN 978-0-9940507-0-0 (paperback)

1. Mineral industries--Canada--Comic books, strips, etc. 2. Gold mines and mining--Guatemala--Comic books, strips, etc. 3. Aluminum mines and mining--India--Comic books, strips, etc. 4. Uranium mines and mining--Québec (Prov-ince)--Comic books, strips, etc. 5. Oil sands industry--Alberta--Comic books, strips, etc.

I. Widgington, David, 1964-, editor II. Dubois, Frédéric, 1977-, editor III. Tessier, Marc, 1962-, editor.

HD9506.C32E98 2015 338.7'62209710222 C2015-904716-1

This book is dedicated to:

the communities resisting illegitimate extraction,
serious, dedicated and independently-minded reporters & investigators,
and
all the artists who put their talent to work for the common good.

EXTRACTION

Comix Reportage

edited by
David Widgington
Frédéric Dubois
Marc Tessier

EXTRACTION

28-40
CAUTION
CORPORATION
KING-BEAVE
ASBESTOS
KB
362
CANADIA
STANDAR
CAUTION
CONTAINS ASBESTOS FIBERS
AVOID CREATING DUST
SERIOUS BODILY HARM
KB-362-3T
CAUTION
CONTAINS ASBESTOS
AVOID CREATING DUST
BREATHING ASBESTOS DUST MAY CAUSE
SERIOUS BODILY HARM

A PARCEL OF FACTS WRAPPED IN COMIX

When Ad Astra Comix first approached me as publisher of *Extraction!*, to republish, I was elated. The first edition sold out quickly after its initial publication in 2007 and a second print run was improbable, particularly after the closure of Cumulus Press in 2008. The issues reported on in the first edition are still relevant and the book's revival would reveal the remarkable work injected into the project by the collective of illustrators, comics artists, journalists and editors. The combination of journalism and comics was an experiment that we all undertook with enthusiasm.

The world is now in a phase of extreme extraction, or the removal from the ground of minerals and fossil carbon that were previously too expensive or technologically impossible to reach. Capitalism's insatiable appetite for perpetual growth has pushed the extractive industry to seek increased profits in previously inaccessible areas.

Shale gas fracking on the Gaspé Peninsula and Île d'Anticosti, open-pit mines in the center of the town of Malartic and aggressive tar sands expansion are but a few examples of this practice of extreme extraction in Canada that have serious consequences on people's lives, on communities' abilities to decide on their own futures and on natural environments that inevitably get spoiled for generations.

It has been five years since the Deepwater Horizon explosion that leaked an astounding five million barrels of oil into the Gulf of Mexico. A recent report from the American Academy of Sciences states that two million barrels of the toxic sludge still sit on the bottom of the Gulf! The April 8, 2015 oil spill in Vancouver's English Bay has led Fisheries and Oceans Canada to close the city's Burrard Inlet to all recreational fishing for shellfish and ground fish. The transport of oil by train has also proven unsafe since the train wreck and subsequent explosion in the town centre of Lac Mégantic that destroyed half of the downtown core, killing 47 people. The March 2011 Fukushima Nuclear disaster released 300 million metric tons of highly radioactive water into the Pacific Ocean. In February 2014, scientists reported at the American Geophysical Union's Ocean Sciences that a radioactive plume from Fukushima was measured in monitoring stations off the coast of Vancouver.

The genesis of the project

It was on a trip to Berlin in January 2007 that Extraction! was first conceptualized. I was visiting co-editor Frédéric Dubois, squatting his Kreuzberg flat. During a late-night conversation over a bottle of Czech beer and a paper bag of mixed nuts from the Turkish shop down the street, we discussed Canadian mining projects around the world. Many of these projects are vehemently opposed by local communities, yet permits are given by national/regional governments, people get displaced, and Canadian multinationals gain access to the mineral wealth.

Now, the planned expansion and aggressive federal government promotion of tar sands extraction — and the resistance to it — has had an influence in our desire to republish *Extraction!*. The thousands of kilometres of oil pipelines required to export the bitumen would pass through communities that don't want them, and across hundreds of rivers that supply drinking water to densely populated cities.

Both Frédéric and I had previously reported on Canadian mining activity around the world. We asked ourselves questions. How can there be so little critical media attention paid to an industry with so many documented cases of opposition by communities negatively affected by Canadian mining practices? How does the West's insatiable appetite for resources affect those living on the land under which the resources are found? What could we do to bring attention to these issues? Since the book's initial publication, The Guardian newspaper's "Keep It In the Ground" project seeks to shame the Gates Foundation and Wellcome Trust, the world's two largest foundations, into divesting from fossil fuels. The Pulitzer Prize-winning newspaper is dedicating journalistic resources to provide critical reporting about the extraction of fossil carbon and its role in global warming. This is a positive shift in journalistic practices by a major international newspaper.

Extraction! is for an audience that does not question Canadian mythology that emphasizes its gentle non-aggressive nature as United Nations peace-keeper or as a provider of humanitarian aid to less developed countries. It is for a readership that is irresponsive to the unflinching government support of the extractive industry. Since the 2013 dissolution of the Canadian International Development Agency into the department of Foreign Affairs and International Trade, the government has provided funding for Canadian NGOs who partner with Canadian mining corporations to implement corporate social responsibility using "development" money to white-wash Canadian mining corporations' environmental and human rights abuses in developing countries. These comics reportages are for a viewership that is unaware of Canada's vote against the United Nations declaration on aboriginal rights or its unilateral decision to pull out of the Kyoto Protocol.

Alain Denault and others have written extensively about Canada as a mining industry tax haven, where a majority of world mining corporations are listed on the Toronto Stock Exchange to benefit from our century-old mining laws. The extractive corporations benefit from financial investment from the Canadian Pension Plan, to which every employee in Canada contributes. It is important that critical perspectives of Canadian mining practices reach audiences of dividend seekers, union workers, and current pensioners to clarify their role as investors in an industry that violently removes people from their homeland, in an industry that usually leaves toxic landscapes behind after the ore is gone, and in an industry that receives huge government subsidies to guarantee increased corporate profits.

After our Berlin discussions, we were set: a book of comics journalism about Canadian mining practices in different parts of the world would be published. Now nearly a decade later, the second edition reveals that interest in unsustainable mining practices remains.

Combining comics with journalism

In 2007, the Centre Pompidou in Paris hosted Rencontre BD reporters, a conference within an exhibit about comics journalism that revolved around the common theme of travel and issues relating to the observation of others, the selection of what to transcribe, the subjective representation of reality and its relationship with other visual forms of reporting: photojournalism or documentary filmmaking. Travel was central to the exhibit because journalism relies on a certain amount of fieldwork to access all of the information needed to get to the bottom of an issue: conducting interviews, observing the scene, taking notes, sketching details. All good journalists get key information firsthand from original sources that are most often near to where the subject is located.

In comics reportages like those by Joe Sacco, where writer and illustrator are the same individual, the illustration portion of the process of making comics is based on memory. Memory from reminders formulated during the weeks and months it takes to visually articulate the information gathered during fieldwork. Photographs, notes, video footage, sketches, audio recordings are all used to compose the illustrations, delineate the panels, develop the narrative.

In the case of Extraction!, the journalist and comics artist of each report are two separate people, matched for the duration of the book Three of the four artists have no memories from the fieldwork because they had not actually gone into the field with the reporters. Joe Ollmann, Phil Angers and Stanley Wany illustrated their comics solely from visual cues without having experienced the context within which the cues emerged and without having gone on location where the extractive projects are situated. Ruth Tait is the exception, having gone to Mont-Laurier with her journalist counterpart, Sophie Toupin, to investigate the uranium exploration in the Upper Laurentians, three hours north of Montréal.

The success of these comics reportages relies on the effective combination of journalistic reports with sequential art, despite their inadvertent rivalry. Here the challenge persisted between the 'give' of verifiable facts and the 'take' of graphic interpretation, between the 'push' of fact-based details and the 'pull' of a sequential narrative. Their strength is due not from the combination of the text and image, content and structure but actually from the added meaning derived from the interaction between the symbolic and the realistic, the literal and the figurative. The cartoon characters inserted into the chapter on bitumen extraction in Alberta uses figurative symbolism to insinuate meaning absent in the script but present in the original journalistic report. How does caricature affect journalistic integrity by adding editorial bias into a report containing verifiable facts?

During the transition to narrative script, characters emerged, scenes were set, and the chronology determined. In each of the three scripts written by the reporters, the journalists themselves became the narrative characters who threaded their way through the stories, making them actual thinking feeling human beings rather than unseen narrators of information. In the tar sands script, based on Petr Cizek's report and written by Marc Tessier, the journalist is absent from the comics, replaced with a fictional narrator on a soapbox who transmits the information to a gathering crowd. How does the fictional character compromise the journalism of the initial report if he is the conduit of the journalist's investigation?

How far from accurate visual representation can a comics artist take factual information for the resulting comics to assert itself as reportage? At which point does it become fiction?

When a comics artist and a journalist collaborate toward a single outcome, there are bound to be compromises from both sides. Occasionally, the scripting phase imposed a *mise en scène*, which blurs the boundaries between fact and fiction, but the essence was never compromised.

This approach to comics journalism, which divided the tasks of journalist and comics artist, raised many questions and provided significant challenges. The resulting four comics reportages met — even surpassed — our expectations. It is a graphic novel that investigates an industry thriving on government grants, self-regulation, corporate takeovers, stock market fluctuations, and the race to control the world's remaining geological resources. We have a comic book that began with serious journalism: verifiable facts, trusted sources, investigation and analysis, and ended disarmingly wrapped in comics. A secret weapon for the uninitiated who may stumble onto it by the sheer beauty of the artwork and strength of the narratives.

Comic books have reputations of being unpretentious, visually appealing and easy to read. Such a mix may provide the reader/viewer with a critical glimpse at global extraction practices that they may otherwise never come across. This is *Extraction!*'s *raison d'être*.

David Widgington
Montréal, June
2015

A BLITZKRIEG CALLED EXTRACTION

The extraction of natural resources in today's capitalism is a dirty business. Since the turn of the millennium, most energy and mineral prices have skyrocketed. Junior mining adventurers and unscrupulous corporate oligopolies rush into new territory to suck what's hot out of the earth's lucrative veins. But that alone hardly qualifies as news.

In recent history, the war for resources was represented as colonization. Now that most former colonies are sovereign nations (at least on paper), competition for resources grows at an alarming rate.

Canada and Australia's extractive industries developed phenomenally after the 1950s. More recently, China's state enterprises have multiplied investments in Africa and Latin America. Brazil's giant, CVRD, has taken over companies at an accelerated rate. Russia's Gazprom worked its way deep into continental Europe.

What is new in today's gas, oil and mining industries, is the pace at which exploration, extraction, transformation and delivery is taking place. The German magazine *Der Spiegel* speaks of a Third World War for the world's resources. But it is more of a *Blitzkrieg*.

The precedent-setting magnitude of extractive projects now rivals any megalomaniac's wildest dreams. In times when open-pit mines have become the norm, even the displacement of entire towns and villages by extractors no longer raises eyebrows or makes for juicy headlines.

Hasty development sponsored by the stock markets in New York, Toronto and Frankfurt — to name just a few — is a key component of this race for resources.

More fundamental and less cyclical though, lies the dogma of economic growth, which encourages disloyal corporate practices and oligopoly-friendly policies on behalf of governments. Most politicians, the business community and scores of academics insist on growth even when the population in most industrialized countries is in decline and over-consumption is at an all time high.

In the last decade, the fusion of the economic growth mantra and the high value of some natural resources became the driving force behind what is increasingly seen by the public as irresponsible extraction.

Despite the fact that 'corporate social responsibility' (CSR) has now become a trendy expression among top industry bureaucrats, mining chief executives, engineering academics and human rights organizations, many extractive projects are setting precedents of large-scale ecosystem disruption and of human dignity violations. Size and speed are good indicators of an industrial sector running amok.

'Mining communities' most often live in an export-driven drama where the extractor is the main benefactor. After only twenty years of intensive exploitation, the mining corporation exits, often leaving a devastated landscape behind. It also leaves behind 'externalities' — long term expenses unaccounted for by the exploiter — that the local taxpayers will need to pay for. In Québec alone, the government has calculated 700 abandoned mining sites and will spend $1.2 billion of public funds to restore them. What's worse is that belligerent policies of most of the industrial nations are resource-driven.

Economic and military power regularly join forces to access and exploit resources, which disturbs communities who happen to live in the valleys and on the mountains containing valuable mineral deposits. *Extraction!* presents varying examples of an extractive *Blitzkrieg*. Three out of the four reportages expose the ramifications of having military, paramilitary and other non-civil institutions involved in the decisions of how resources are accessed, extracted or used.

Canada as the top miner

The book you are holding in your hands is a modest contribution to an existing tradition of reporting on mining. It comes at a time when citizens of many countries are increasingly worried about mining practices and their impact on the environment. They are worried about jobs and their livelihood, as well as long-term revenues from mining.

Canada is a haven to more mining corporations than all other countries combined. Extractive industries drive the Canadian economy. 60% of investments in extractive activities worldwide are reputed to be Canadian or from a corporation registered on one of Canada's stock exchanges. It creates significant wealth for shareholders including the Canadian Pension Plan, to which every employee in Canada contributes.

Canada is the biggest oil supplier to the world's first economy (that of the USA). The province of Alberta's Tar Sands production exceeds 1,9 million barrels per day and could reach 3.8 million barrels per day by 2022. The Tar Sands represent the largest single oil deposit in the world at 1.6 trillion barrels. To date, only a small portion — 175 billion barrels — has been categorized as proven reserves, but already, Canada is reputed to possess the second largest oil reserves in the world after Saudi Arabia.

Canadian mining benefits from a long history of government subsidies. Access to start-up capital funding through the stock market is relatively easy. Royalties are low in many mining sectors, mining legislation is highly favourable to large-scale mining and the technologies used in the Canadian mining sector are state-of-the-art.

Human resources are abundant and their expertise is unmatched. Every second engineer or manager you'll meet in the fields of extraction has a Canadian passport.

A made-in-Canada conquest

The irony of these glamorous facts is that Canada's hold on its 'gold medal' as the world's top miner is starting to erode Canada's reputation abroad.

Historically seen as a soft power, a peaceful and humanitarian place, Canada is increasingly synonymous with evil and destruction. In the last two decades, the media has disseminated exuberant national pride based on Canada's ranking in the top ten among 175 countries in the United Nations Quality of Life Index.

There is little media mention of how Canadians came about the wealth that determines their quality of life nor how the country is experiencing increasing levels of childhood poverty. International criticism for its treatment of indigenous communities is difficult to find in the media, as is the country's increasing divide between the wealthy and the poor.

The Canadian myth that emphasizes a gentle non-aggressive citizenry who proudly refers to the country's historic military role as United Nations peacekeeper and a human rights advocate is... a myth.

Many of the mining projects Canadian corporations are involved in, are vehemently opposed by local communities, yet permits are given by national/regional governments, people get displaced, and Canadian transnationals gain access to the mineral wealth.

The Canadian government decided to establish the National Roundtable Discussions on Corporate Social Responsibility and the Extractive Industries back in 2006-2007 to hop on the 'social responsibility' bandwagon and improve its public profile. There were four public assemblies. Members of civil society, the private sector and the government met in Vancouver, Calgary, Toronto and Montréal to discuss what extractive corporations should be responsible for.

The recommendations that came out of this roundtable process could have marked a new step in addressing the increasing human rights and ecology abuses left in the wake of extractive tractors and sludge-hammers. But they fail to do so! They fall short of advocating international extraction regulation and tough checks and balances. Instead they demand the creation of an ombud's office and clear criteria to orient responsible extraction. This is not enough, particularly as one looks at several multinationals' abuse records. This public relations move, at times a genuine attempt to consult with extraction experts, was meant to calm down criticism about mining, gas and oil exploitation.

Goldcorp's documented pollution of rivers in Honduras; Gabriel Resources' avid pursuit for a licence to dig gold in the hills of Transylvania with an invalid Environmental Impact Assessment; and Canada's junior exploration firms taking advantage of lean legislation in African nations for uranium are examples that shout for tough and rigorous laws on extraction — in Canada and internationally. These examples also mean that more needs to be done to document and publicize the industry to develop awareness within the larger public.

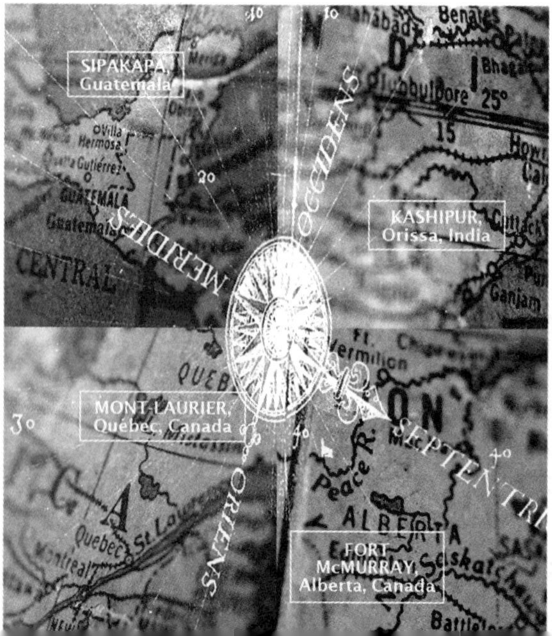

Some increasingly argue that mining is not just a Canadian phenomenon. Capital interests from other countries are buying up more and more Canadian corporations. Instead of focusing on the many negative impacts of mining, they argue, we should be more patriotic and support "our" companies. The truth is that "our" companies are completely out of control. The general public has little influence on the course taken by a corporation, be it Canadian or from another jurisdiction.

The made-in-Canada *Blitzkrieg* in extraction is real and it is being raged upon communities with the support of our public policies, sponsored in part by our money and with the help of our ignorance of the facts. Our patriotism lies with the common good. This is why community and public interest were our points of reference to know whether the way resources extraction is done can be regarded as responsible or not.

It is important that critical perspectives on mining practices reach audiences of dividend seekers, union workers, and future pensioners, among others. This, in parallel with stringent legislation, will allow communities and those resisting mining projects to understand and prevent the green washing tactics of self-proclaimed 'good corporate citizens'.

Four resources, four reportages

The story about mining cannot be considered as a black-and-white issue. The extraction process is complex, requiring nuances in the discourse. There are well-intentioned mining entrepreneurs, just as there are opportunistic and corrupt non-governmental organizations. There are self-interested civil servants and pro-mining indigenous leaders. Our interest is therefore not to take a position against the sector as such, but rather to take a hard look at the Canadian-led *Blitzkrieg* and to listen to what dissenting voices have to say.

But nuances are rarely explicit, so journalism needs to provide an overview of mining's realities by documenting, analyzing and critically assessing the impacts mining has on communities and on the environment. That's the responsibility of critical, independent and honest reporters.

Extraction! delves into the implications of Canadian mining practices. It considers the communities affected by them. It is certainly not anti-development, but it adopts a critical and journalistic view of all actors involved in some significant tragedies of our time.

This book of comics reportage is about the exploration, exploitation and extraction of oil, uranium, bauxite and gold. Although we could have chosen to look into more than these four resources, the intention is to initiate a comics reportage series — with this as its first book.

We chose gold, uranium, bauxite and oil for a number of reasons. We aimed for the news value and their significance in depicting the extractive sectors that are important at the time. We also wanted to work with those journalists who had experience reporting on mining and who were already familiar with the particular mineral.

Gold was a natural fit, since it is the resource for which Canada's corporations are notorious. Self-serving behaviour, a complete disrespect for local communities and bad faith are legendary in the gold sector. In recent years, with gold selling at high prices, many abandoned mines have been reactivated. The pressure from developments such as in Rosia Montana, Romania, has gone as far as igniting conflicts with neighbouring Hungary.

Dawn Paley's outstanding and courageous journalism in Guatemala delves deeper into the gold file. She investigates Goldcorp's presence in indigenous communities and brings the news back to Vancouver, to the corporation's corridors of power.

Eight years later, Goldcorp continues to operate the Marlin mine in the department of San Marcos, Guatemala. By all accounts, social conflict has risen, and the negative environmental impacts of the project are increasingly manifest and continue to be strongly contested by villagers in Sipakapa and San Miguel Ixtahuacán. One of the most visible opponents of the mine, Diadora Hernández, was shot in the head in 2010, but she miraculously survived. Doña Elida, who inspired the character of Luisa in the comic, passed away in 2012. Elida's kindness and spirit of struggle remain present for all of us who knew her. Also in 2012, Bishop Ramazzini was moved to the department of Huehuetenango, where he continues to support struggles against extraction.

Uranium was impossible to ignore. This mineral represents the future, as the Kyoto Protocol is being implemented internationally, and nuclear energy is touted as a green alternative to dirty oil. Canada's uranium history as a source for the bombs dropped on Hiroshima and Nagasaki was worth pointing out, to debate the sustainability and ethics of its use.

Journalist Sophie Toupin's interview skills and sharp research helped her access a small-town uranium story with universal implications. In an independent and devoted manner, she demonstrated where that sector is headed in the region and around the world.

The Fukushima catastrophe of 2011, the largest nuclear incident since the 1986 Chernobyl disaster, has revived the anti-nuclear movement in many countries. However, despite this catastrophe, the uranium mining industry and the pro-nuclear lobby groups are still arguing that nuclear energy is an answer to global warming. In her most recent book, *This Changes Everything* (2014), Naomi Klein notes that 12 percent of the world's power is supplied by nuclear energy from obsolete facilities that are becoming more and more dangerous, while arguing that governments should transition from high-risk energy like nuclear to renewable energy such as water, air and sun. She also uncloaks the argument used by the pro-nuclear industry that nuclear power is releasing less CO_2 in the atmosphere and therefore a response to global warming, when one looks at the "vast amounts of fossil fuels it takes to mine, transport and enrich uranium" (2014, p. 120).

In Québec, where Toupin's reportage story took place in 2007, l'Association de protection de l'environnement des Hautes-Laurentides (L'APEHL) is still fighting for a permanent moratorium against the exploration, exploitation and extraction of uranium in the province of Québec. Like Klein, l'APEHL is consistently trying to demystify the green washing argument used by the industry in an attempt for the Québec government to lift the temporary moratorium against uranium mining put in place in 2013. As a result of the temporary moratorium, the company Strateco Resources Inc., which had started a uranium project on Cree land in the northern part of Québec under former premier Jean Charest's Plan Nord is suing the Québec government for $190-million in investment losses.

Third, we chose bauxite because Alcan (now Rio Tinto Alcan), one of the world's leading aluminium producers, is based in Montréal. Our intimate knowledge of Alcan's involvement in a bauxite extraction and smelting case in East-India, with its strong opposition from local activists, was sufficient to convince us of its relevance.

International reporter Tamara Herman, who had visited the mining co ncession area two years earlier, has reactivated the file. She double-checked the received truths from both sides of the spectrum and reconstituted the facts along the lines of common-good reporting.

In 2007, Alcan pulled out of the Utkal project, stating that the decision was 'in the best interest of all our stakeholders.' Alcan sold its share to Hindalco Industries, its former Indian partner. With money changing hands and the mine proposal continuing on, this so-called victory in Canada meant little for the Adivasi people in Kashipur. Communities continued to resist the mine until production began in 2013, while battles over compensation for lost land and water shortages still rage.

Extraction! would not have been as nuanced and comprehensive as it is, were it not for the mother of all resources: oil. Carbon-dependent Canada and the United States, corporate oil giants, NGOs and resource think tanks are lining up for an important showdown in the tar sands of Alberta. There are a thousand reportages to be made about oil and we necessarily had to limit ourselves to one.

Petr Cizek, a seasoned northerner, cartographer and writer, caught our attention with a fine analysis and well-researched article about the liquid relationships between the various players of Canada's oil bonanza.

The expansion of tar sands bitumen extraction is planned to double by 2020 and even triple by 2035. The proposed pipelines needed to export the bitumen (Enbridge's western Northern Gateway, its eastern Line 9 and TransCanada's southbound Keystone XL, eastbound Energy East) are being successfully opposed by activists across the continent. The recent decline in world oil prices have reduced spending in the tar sands and have delayed expansion plans.

The limits of *Extraction!*

The value of *Extraction!* lies in the depth of the reporting, the quality of the illustrations and its commitment to keep each of the extraction stories accessible to a general readership.

There are shortcomings in this book that deserve highlighting for reasons of transparency and fairness. The four chapters that make up Extraction! fall short of describing certain fundamentals about the industry.

An important blind spot is geographical: Africa is absent. This was unintentional.

We had started investigating several diamond extraction cases in African countries. The idea was to move beyond the 'blood diamond' narratives of the last decade and demystify the official and illegal mining dynamics practiced today.

We decided not to elaborate on the escalation of 'counter-insurgency tactics' used by mining corporations to delegitimize their opponents. There are stories we could have told about journalists producing pro-industry documentary films, such as *Mine Your Own Business*, a propaganda piece taking aim at environmentalists. We would have insisted on company-floated NGOs and the whitewashing of corporate images with the hiring of former prime ministers of Canada as lobbyists, but space restraints disallowed for it in this volume.

Living conditions of migrant workers, of rank-and-file miners and the security of their workplaces are not put to the fore in *Extraction!*. We believe that labour issues have historically been the most striking illustrations of the large-scale abuse that the extractive sector is capable of. Today, many small-scale and cooperative miners — and miners in general — face risks to their health, which are often life-threatening.

In most Canadian corporations though, it must be recognized that labour standards are relatively high and our journalists have seen the urgency as being rooted in other dimensions of the mining business. We have deliberately decided not to put the accent on labour issues. In the meantime, we're receiving weekly reminders via newswires, about coal mine accidents, from China to Chile, and back.

Renewable energies drawn from wind, water, sunlight and geothermal heat are absent from most of the coverage herein. Clever technological advances using biomass and biofuels for transportation would certainly have broadened our understanding of the alternatives to conflict-inducing extraction. After the resistance, the alternatives.

There are certainly many other aspects that would merit media attention and in-depth analysis. We have had to compose with very limited means and a tight timeframe. Our hope is that the following chapters do contribute in a constructive way to the debate around extraction.

The reaction *Extraction!* is trying to trigger is best summed up by humanist photographer Sebastião Salgado: "Are we condemned to be largely spectators? Can we affect the course of events? Can we claim 'compassion fatigue' when we show no sign of consumption fatigue?"

We hope that *Extraction!*, with its Canadian angle, fresh reporting and accessible style, will inspire debate. We also hope that the stories of resistance that we celebrate herein will be understood as acts of courage and survival: fundamental to human development. There are many ways to affect the course of events. One is by extracting stories and sharing them. This is what we have done.

Frédéric
Dubois Berlin,
June 2015

GOLD

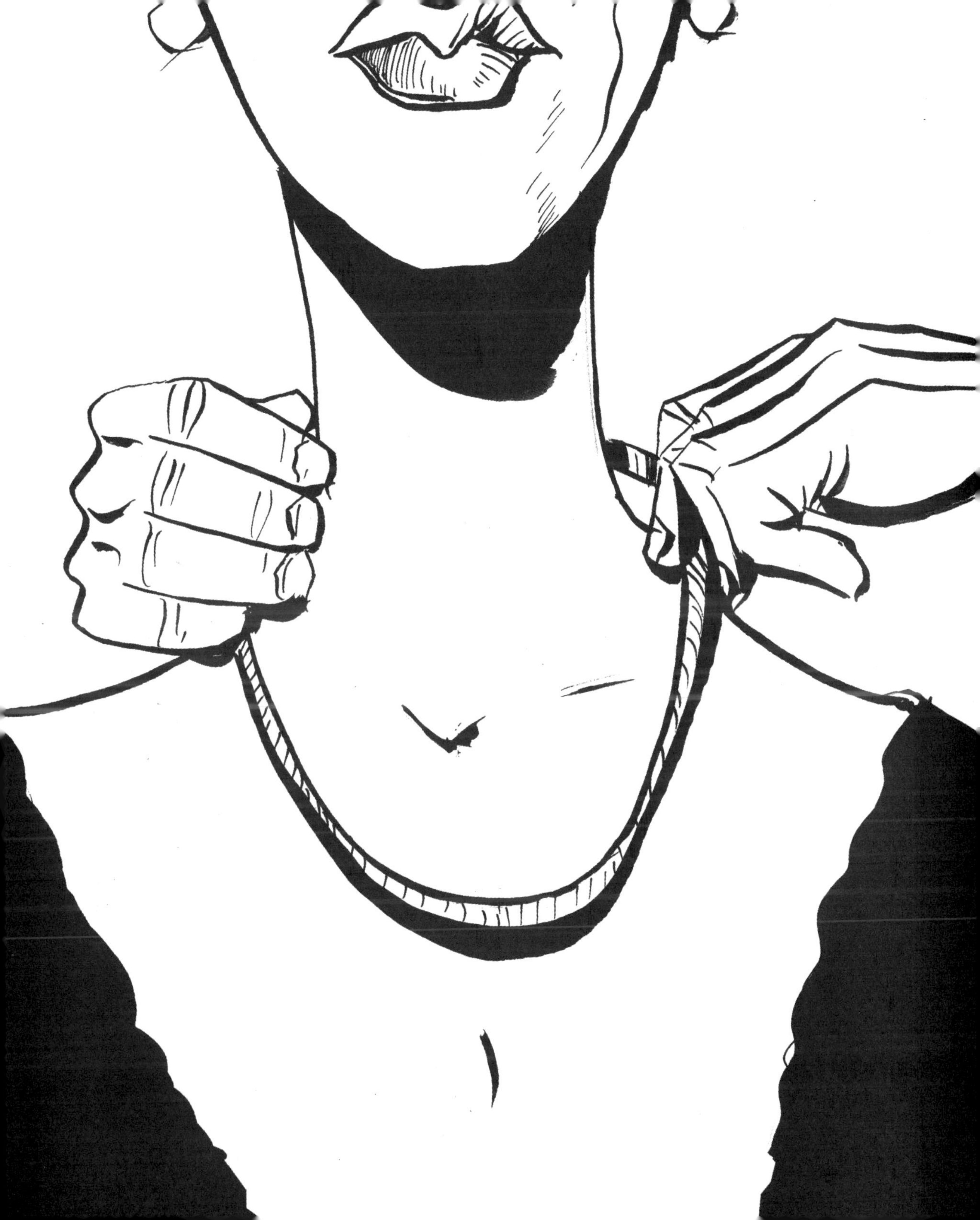

GOLD

Taking the Heart from the Land

REPORTAGE BY DAWN PALEY
COMIX BY JOE OLLMANN

GOLD

NOTE : Some names have been changed to protect the subjects.

VANCOUVER, EARLY NEXT MORNING.
HE'S GOING TO KILL US!
GUATEMALA CITY.
HE'S GOING TO KILL US!
TODAY I ARRIVED IN GUATEMALA'S CAPITAL...
TAXI
SO CLOSE TO HOME IN A WAY...
ROBERTO
BUT IN ANOTHER WAY, REAL DAMN FAR...

BY A STROKE OF LUCK, I'M MEETING WITH LUIS SOLANO THIS MORNING. HE'S A JOURNALIST WHO'S BEEN WRITING ABOUT MINING ISSUES IN GUATEMALA FOR THE LAST TWENTY YEARS.
HOTEL PAN AM
UN CAFECITO, POR FAVOR.
HOLA, DAWN.
UN CAFECITO, POR FAVOR.
HOLA.
CAN YOU GIVE ME AN OVERVIEW WITH REGARDS TO MINING IN GUATEMALA?
WELL, THE FIRST NORTH AMERICAN PROJECT IN GUATEMALA WAS AN INCO PROJECT THAT WE CALLED EXMIBAL.
THE FIRST NICKEL MINING CONCESSIONS WERE GRANTED IN THE LATE 1950s, RIGHT AFTER THE COUP D'ÉTAT AGAINST PRESIDENT JACOBO ARBENZ.
ARBENZ
THE COUP D'ÉTAT WAS ENGINEERED BY THE CIA TO BENEFIT THE UNITED FRUIT COMPANY, A U.S. BANANA COMPANY THAT WOULD HAVE LOST SOME UNUSED LAND THROUGH ARBENZ'S LAND REFORM/REDISTRIBUTION PLAN.
MINING LAWS WERE REWRITTEN IN THE 1960s BY PEOPLE ON THE INCO PAYROLL TO FACILITATE THE PROJECT.
POLITICIANS WHO OPPOSED THE LAW CHANGES WERE REMOVED FROM OFFICE. SOME WERE KILLED. THE MINE WAS ONLY IN OPERATION FOR A COUPLE OF YEARS.

MANY PEOPLE WERE KILLED – LOCALS WHO RESISTED THE PROJECT OR EVEN SIMPLY ASSERTED THEIR HISTORICAL RIGHTS TO THE LAND...
AS WELL AS CONGRESSMEN AND LAWYERS WHO ARGUED IT WOULDN'T BENEFIT GUATEMALA...
LECHE
–OTHERS DISAPPEARED.
DESAPARECIDO
PABLO BAC CAAL
RELATIVES OF THE VICTIMS OF THESE CRIMES STILL LIVE IN THE AREA. THERE ARE A HANDFUL OF CANADIAN COMPANIES AROUND THERE EXPLORING FOR NICKEL AGAIN NOW THAT PRICES ARE UP AND THE INTERNAL CONFLICT IS OVER.
EXMIBA
NO TRESPAS
THE PEACE PROCESS WAS SEEN BY THE GOVERNMENT OF GUATEMALA AS NECESSARY TO ENCOURAGE FOREIGN INVESTMENT.
AFTER 36 YEARS OF INTERNAL CONFLICT IN WHICH MORE THAN 200,000 MAINLY INDIGENOUS PEOPLE WERE KILLED, THE LAST PEACE ACCORD WAS SIGNED IN DECEMBER, 1996. IN 1997, THE MINING CODE WAS CHANGED. ROYALTIES WERE REDUCED FROM 7% TO 1%.

THERE ARE INTERNATIONAL AGREE-MENTS THAT SAY THAT INDIGENOUS PEOPLE SHOULD BE CONSULTED BEFORE DEVELOPMENT TAKES PLACE ON THEIR LAND. ONE OF THESE IS ILO 169.
WHAT'S ILO 169?
ILO 169 IS THE U.N. CONVENTION ON INDIGENOUS AND TRIBAL PEOPLES. GUATEMALA RATIFIED IT, MEANING THAT IT IS BOUND TO CONSULT IN-DIGENOUS COMMUNITIES BEFORE EMITTING DEVELOPMENT PERMITS.
YOU'LL SEE FOR YOURSELF WHERE THAT IS AT WHEN YOU VISIT THE COMMUNITIES AROUND THE MARLIN MINE... -DON'T FORGET TO TAKE A COPY OF MY BOOK.
GRACIAS LUIS, YOU'VE GIVEN ME LOTS TO THINK ABOUT.
YOU'RE WELCOME.
OH GREAT, 12 PAGES OF FOOTNOTES!?!
THIS IS THE KIND OF STORY IT WOULD TAKE YEARS TO RESEARCH.
ANANAS!
IT'S GOING TO BE TOUGH TO DO IT JUSTICE.
CARNE
internet
10
Am in Guate. Every-thing is fine. Hope-fully things will go well in San Marcos...

I TRAVELED FROM GUATEMALA CITY TO SAN MARCOS, THROUGH THE HIGH-LANDS ON A "CHICKEN" BUS.
CERVEZA
THE RIDE IS LONGER AND FARTHER (AND SLOWER! AND BUMPIER!) THAN I EXPECTED.
ALL THAT SLEEPING ON THE BUS IS INCREDIBLY TIRING...
BUS
I'VE GOT A CONTACT HERE FOR JORGE GRANADA FROM THE CAMPESINO WORKERS' MOVE-MENT. MIGHT BE A GOOD PLACE TO START.
HOTEL
TV
OKAY, HASTA 10:30.
SAN MARCOS IS SMALLER, COOLER, AND MUCH CALMER THAN GUATEMALA CITY.
MTC
BUENOS DIAS, JORGE?
AHORA VENGO!
NO SE VENDE!
JORGE SHOULD BE ABLE TO GIVE A GOOD OVERVIEW. HE WORKS CLOSELY WITH COMMUNITIES.

TO START OFF WITH: WHAT CAN YOU TELL ME ABOUT THE MARLIN MINE?
THE MARLIN MINE IS AN OPEN PIT GOLD MINE. IT'S IN LANDS BELONGING TO TWO MUNICIPALITIES: SAN MIGUEL IXTAHUACÁN AND SIPAKAPA, WHICH ARE, RESPECTIVELY MAYAN MAM AND MAYAN SIPAKAPENSE TERRITORY.
HOW DID THE MINING COMPANY ENTER THE AREA?
PEOPLE LIVING AROUND WHERE THE MINE IS NOW, FELT PRESSURED TO SELL THEIR LANDS EARLY ON BEFORE REALIZING THAT A MINE WAS BEING BUILT. CONSTRUC-TION STARTED IN 2004, AROUND THE SAME TIME AS CANADIAN AMBASSADOR TO GUATEMALA, JAMES LAMBERT, WROTE A PRO-MINING EDITORIAL IN PRENSA LIBRE.
THE MINING COMPANY IS VANCOUVER-BASED GOLD CORP, WHICH AQUIRED GLAMIS GOLD IN 2006. THE FULLY OWNED SUBSIDIARY IS CALLED MONTANA EXPLORADORA DE GUAT-EMALA. ALL CANADIANS HAVE INVESTED IN GOLDCORP THROUGH THE CANADIAN PENSION PLAN TO THE AMOUNT OF $181 MILLION.
MONTANA EXPLORADORA DE GUATEMALA
GLAMIS GOLD
GOLD CORP
CANADA PENSION PLAN
CAN YOU TALK ABOUT THE RES-ISTANCE TO THIS MINING PROJECT?
THERE'S BEEN RES-ISTANCE IN SAN MIGUEL AND IN SIPAKAPA, BUT THE PEOPLE IN SAN MIGUEL HAVE FACED MORE REPRESSION AND IT HAS BEEN HARDER FOR THEM TO ORGANIZE.
A 2004 SURVEY INDICATED THAT 95% OF PEOPLE LIVING NEAR THE MARLIN MINE, MOST OF WHOM ARE INDIGENOUS, OPPOSED THE MINE. THERE HAS BEEN A LOT OF ATTENTION ON THE PROJECT BY HUMAN RIGHTS ACTIVISTS, NGOs, AND JOURNALISTS, SINCE IT RECEIVED A $45,000,000 LOAN FROM THE IN-TERNATIONAL FINANCE CORPORATION (IFC) THE PRIVATE SECTOR ARM OF THE WORLD BANK.
45 MILLION DOLLARS
IT'S VERY CLEAR IN THIS CASE, AS IN ALL OF GUATEMALA: THE MAP OF POVERTY, THE MAP OF INDIGENOUS PEOPLES AND THE MAP OF MINING PROJECTS LINE UP PERFECTLY.
MINE
AREA OF INDIGENOUS POVERTY.

ANOTHER THREE HOURS ON THE BUS TO SIPAKAPA.
BUS
WHAT'S THAT, OVER THERE?
THAT'S THE MINE, HAVEN'T YOU HEARD? IT'S THE BIG-GEST NEWS IN TOWN!

CANTINA
HOLA, LUISA.
BENVENIDOS.
SIPAKAPA IS HOME TO ABOUT 14,000 PEOPLE, MOST OF WHOM ARE MAYAN SIPAKAPENSE. IT IS DIVIDED INTO 14 VILLAGES, WITH ABOUT 1000 PEOPLE LIVING IN THE CENTRE.
LUISA TAKES ME TO HER HOUSE FROM WHERE SHE RUNS A CAFETER-IA. SHE BRINGS ME COFFEE AND TORTILLAS.
SO, YOU'RE HERE TO WRITE ABOUT THE MINE, ARE YOU?
THAT'S RIGHT.
THE MINE ISN'T BENEFITING US... WE LIKE IT PEACEFUL HERE, WITHOUT TRANSPORT TRUCKS GOING PAST AND ALL OF THE VICES THAT MINING WOULD BRING INTO OUR TOWN.
AND IMAGINE, THE WORLD BANK GIVING THE COMPANY $45 MILLION AS IF IT WAS FOR OUR DEVELOPMENT!
WHAT WE COULD DO WITH $45 MILLION...
IT'S THE FIRST LOAN OF ITS KIND SINCE 2001, WHEN THE IFC STARTED A REVIEW PRO-CESS FOR LOAN POLICY TO THE EXTRACTIVE INDUSTRIES.

HOLA.
HOLA, HECTOR.
OKAY, FIRST TELL ME ABOUT THE CONSULTA.
IT STARTS WITH WHAT HAPPENED.
... IN SOLOLÁ IN JANUARY OF 2005. THE PEOPLE THERE WERE BLOCK-ING THE PASSAGE OF A CYLINDER NEEDED FOR USE AT THE MINE.
POLICIA
ON JANUARY 11TH, WE LEARNED THAT SOMEONE HAD BEEN KILLED IN THE BLOCKADE AND THAT THE MILITARY AND POLICE WERE ESCORTING THE CYLINDER ALONG THE HIGHWAY.
AFTER 40 DAYS OF BLOCKADING, THE POLICE MOVED IN.
41
SO THE PEOPLE OF SIPAKAPA GOT TOGETHER TO FIGURE OUT HOW WE WOULD ACT. — A PUBLIC MEETING WAS CALLED.
WE WANT THE MUNICIPALITY TO AUTHORIZE A CONSULTA.
MAYOR OF SIPAKAPA.
YES! SI! WE DEMAND A CONSULTA!!
THE CONSULTA WILL ASK WHETHER THE PEOPLE OF SIPAKAPA ARE FOR OR AGAINST MINING.
p-diddy
OKAY, A CONSULTA SHALL BE HELD.

IT WAS DECIDED THAT THE CONSULTA WOULD TAKE PLACE JUNE 18, 2005.
AND WHAT IS THE LEGAL STATUS OF THE CONSULTA?
THE LEGAL FRAMEWORK FOR THE CON-SULTA IS BASED IN THE MUNICIPAL CODE, AND IN OUR RIGHTS AS IN-DIGENOUS PEOPLES, AS DEFINED BY ILO 169.
WHEN THEY FOUND OUT ABOUT OUR PLANS, BUT BEFORE THE CONSULTA WAS CARRIED OUT, MONTANA FILED TWO COMPLAINTS, ONE WITH THE CONSTITUTIONAL COURT AND ANOTHER WITH A LOWER LEVEL COURT, CLAIMING THAT THE PROCESS IS ILLEGAL.
THAT'S STRANGE, SINCE MONTANA'S PUBLIC DOCUMENTS CLAIM THAT PEOPLE IN THE AREA SUPPORT THE MINE.
WE KNEW WE WERE GOING TO GO AHEAD WITH THE CONSULTA, EVEN THOUGH THE COMPANY TRIED TO STOP US.
WE'VE BEEN THE OWNERS OF THESE LANDS FOR THE LAST 5,000 YEARS AND WE HAVE THE RIGHT TO CONTROL WHAT HAPPENS HERE.
THE LOWER LEVEL COURT RULED ON THE SIDE OF THE COMPANY...
ON THE DAY OF THE CONSULTA, THE CONSTITUTIONAL COURT HADN'T RULED, BUT THEY HAD ANNOUNCED IT WAS LEGAL.

IN ALL THIRTEEN COMMUNITIES IN SIPAKAPA, PEOPLE WENT TO VOTE IN THEIR LOCAL HALLS...
REGIST
...USING VARIOUS METHODS TO TALLY THE VOTES.
SI
NO
SI
NO
INK
SI
NO
ALL AGAINST MINING, PLEASE RAISE YOUR HANDS...
ABIERTO
Avon
ALL IN FAVOUR OF MINING, PLEASE RAISE YOUR HANDS...
ABIERTO
Avon.

IN ELEVEN OF THE THIRTEEN COMMUNITIES, THERE WAS A UNANIMOUS NO TO MINING. ONE COMMUNITY WITH A LOW VOTER TURN OUT VOTED YES AND ANOTHER ABSTAINED.
THE MANDATE IS CLEAR. THE PEOPLE OF SIPAKAPA DON'T WANT MINING! PEOPLE WANT TO CONTINUE LIVING FROM THE LAND, BENEFITING FROM THE CLIMATE, THE GOOD SOIL, THE CLEAN AIR AND WATER.
WE'VE ALREADY HAD STUDIES DONE SHOWING CONTAMINATION, AND IT IS ALREADY VERY BAD AT GOLD-CORP'S SAN MARTIN MINE IN HONDURAS.
MEET ME AT THE PARISH HOUSE TO-MORROW FOR A PRESS CONFERENCE. THE BISHOP OF SAN MARCOS, ALVARO RAMAZZINI WILL BE THERE TO ANNOUNCE THE RESULTS.
THE NEXT DAY.
A STUDY SHOWS THAT THERE IS ACID MINE DRAINAGE HAPPENING INTO THE TZÁLA RIVER, DOWN-STREAM FROM THE MINE.
ACID MINE DRAINAGE RESULTS FROM ALL OF THE UPTURNED ROCK AND WASTE FROM THE MINE BEING EX-POSED TO THE ELEMENTS. THE STUDY WAS CARRIED OUT BY FLAVIANO BIANCHINI, AN ITALIAN BIOLOGIST.
WITHOUT AUTHENTIC HUMAN DEVELOP-MENT AND RESPECT FOR NATURE, THERE WON'T BE PEACE IN THIS OR ANY OTHER COUNTRY.

AFTER THE STORY CAME OUT IN THE GUATEMALAN MEDIA...
NEWSPA
MAGAZINES
EL ESPECTADOR
CONTAMINA
IN SAN MARCO
... THE BIOLOGIST WHO CARRIED OUT THE STUDY WAS THREATENED AND FOLLOWED.
#!
HE DECIDED TO GO BACK TO ITALY FOR HIS SAFETY.
TAXI

FROM SIPAKAPA I TOOK A BUS BACK TO GUATEMALA CITY.
TOOK A PLANE AND FLEW HOME.
SO MUCH INFORMATION IN SUCH A SHORT VISIT... HOW TO TIE IT ALL TOGETHER?

I'M ALMOST DONE WITH MY ARTICLE, BUT THERE ARE NEW DEVELOPMENTS EVERY DAY... IT'S HARD TO KEEP UP. WITH IT. I'M INTERVIEWING VINCENTE FICO FROM SIPAKAPA FOR THE LATEST.
WHAT'S HAPPENING THERE, VINCENTE?
IN SAN MIGUEL IXTAHUACÁN, PEOPLE DIDN'T HAVE THE CHANCE TO HOLD A CONSULTA. MOST PEOPLE DIDN'T REAL-IZE WHAT WAS GOING ON UNTIL THE MINERS STARTED BRINGING IN THE HEAVY EQUIPMENT.
AND NOW?
NOW?
THE PEOPLE ERECTED A BLOCKADE ON THE ACCESS ROADS TO THE MINES IN MID-JANUARY, 2007. THEY'VE BEEN OUT THERE FOR ALMOST TWO WEEKS.
NO a lA MINERIA
PEOPLE WERE COMPLAINING THAT THEY WERE UNFAIRLY PAID FOR THEIR PROP-ERTIES. THE COMPANY STILL AC-CESSED THE MINE BY HELICOPTER.
10
YESTERDAY, THERE WAS ANOTHER CONSULTA IN A MUNICIPALITY CALL COMITANCILLO, WHERE PEOPLE AGAIN REJECTED MINING.
CONSULTA COM
DURING THE CONSULTA, THE COPS SWOOPED DOWN IN SAN MIGUEL IXTAHUACAN AND ARRESTED PEOPLE WHO WERE BLOCKADING!
AT THE SAME TIME AS THEY WERE BLOCKADING IN SAN MIGUEL IXTA-HUACÁN, PEOPLE LIVING IN A COM-MUNITY NEAR GOLDCORP'S LOS FILOS MINE IN GUERRERO, MEXICO WERE ALSO BLOCKADING THE ROADS.
POLICIA
POLICIA
THE SITUATION FOR EVERYONE LIVING NEAR THESE PROJECTS IS DIFFICULT. PEOPLE'S HOUSES HAVE STARTED CRACKING, THERE IS NOISE, TRUCKS AND SO ON.

WEEKS GO BY.
rinnG!
WEEKS AFTER I GOT BACK, I GOT A CALL FROM GRAHAME FROM RIGHTS ACTION, A SMALL NGO WITH TWO OFFICES IN CENTRAL AMERICA.
rinnnG!
LOOKS LIKE GOLD CORP'S 2007 AGM IS IN VANCOUVER ON MAY 2. COULD YOU BE THERE WITH HECTOR?
MAY
...SOUNDS... DO-ABLE.
OKAY! RIGHTS ACTION CAN LOOK AFTER THE TICKETS. YOU'LL HAVE TO ORGANIZE A SPEAKING EVENT. LET US KNOW IF WE CAN HELP. GOOD LUCK WITH THE AGM!
ACT
I SPENT THE NEXT TWO MONTHS PLANNING OUT HOW TO BE EFFECTIVE AT THE AGM, AND PREPARING HECTOR'S VISIT. WHICH BROUGHT HIM TO SEVEN CITIES FROM WINNIPEG TO VICTORIA.
...SOME TIME BEFORE MAY 2... THIS MAY...
APRIL

IN NO TIME, I WAS MEETING HECTOR AT THE AIRPORT.
GATE A-4
INTERNATIONAL ARRIVALS
HOLA, HECTOR!
HOLA!
WELCOME!
I ACCOMPANIED HECTOR TO VARIOUS MEETINGS WHERE HE WAS INVITED TO SPEAK ABOUT MINING IN HIS COMMUNITY.
UNA CORPORACIÓN NO ESTA HACIENDO CASO DE LOS LEYES DE UN PAIS SOBERANO...
UH... A COMPANY CAN'T JUST IGNORE... THE.. UH.. LAWS OF ANOTHER LAND...
VANCOUVER, MAY 2. TODAY IS GOING TO BE A BIG TEST... WE'LL SEE HOW GOLD CORP'S SHARE HOLDERS REACT TO OUR QUESTIONS!
YORK ROOM
GOLD CORP AGM
WE HAVE ORDERS NOT TO LET PROTESTERS IN...
GOLD CORP AGM
I HAVE THE LEGAL RIGHT TO ATTEND THIS MEETING. I HAVE A SHAREHOLDER PROXY.
GOLDCORP
share the excitement

... OUR BOARD HAS BEEN EXCELLENT AND HELPED THE COMPANY GROW. ALL IN FAVOUR OF KEEPING OUR BOARD?
Yeah!
CLAP! CLAP!
...MOTION MOVED AND SECONDED.
CLAP CLAP!
YAY!
CLAP CLAP
CLAP! CLAP!
IAN TELFER, CHAIRMAN.
OUR AUDITOR HAS DONE AN EX-CELLENT JOB AUDITING. ALL IN FAVOUR OF KEEPING THE AUDITOR?
CLAP CLAP
MOVED AND SECONDED... WITHOUT FURTHER ADO...
CLAP CLAP
CLAP CLAP
CLAP CLAP
GOLDCO
CLAP CLAP
CLAP CLAP CLAP
CLAP
CLAP CLAP CLAP
THERE ARE VERY SERIOUS HUMAN RIGHTS ABUSES HAPPENING IN 2007 AROUND GOLD CORP MINES IN GUER-RERO, MEXICO AND HONDURAS.
WITH US TODAY IS A GUATEMALAN COMMUNITY MEMBER WHO HAS COME TO SHARE HIS NEGATIVE EXPERIENCES WITH THE COM-PANY WITH YOU...
WORRY NOT! GOLD CORP IS WINNING THE HEART AND MINDS OF THE COMMUNITIES WE OPERATE IN!
KEVIN MCARTHUR, CEO, GOLDCORP.
WE'RE A FIRST-RATE COMPANY.. WE BRING JOBS TO REGIONS WHERE OUR MINING PROJECTS ARE... PROSPERITY!
GOLDCORP HAS PROFITABLE PROJECTS NOT ONLY IN GUAT-EMALA, BUT ALSO IN HONDURAS, CHILE, ARGENTINA...
... SMALL "FRINGE-GROUPS" COMPLAINING..
...SUCCESFUL MINES IN MEXICO, THE DOMINICAN REPUBLIC...

JUST TWO WEEKS AFTER THE AGM, THE CONSTITUTIONAL COURT DECLARED THE CONSULTA LEGAL BUT NON-BINDING, REAFFIRMING KEVIN MCARTHUR'S POSITION...

BUT THE COMMUNITY WILL KEEP STRUGGLING FOR IT'S RIGHT TO CONTROL ITS TERRITORY. — STAY TUNED...

URANIUM

URANIUM

Highway of the Atom

REPORTAGE BY SOPHIE TOUPIN
COMIX BY RUTH TAIT

URANIUM

THANK YOU EVERYONE FOR BEING HERE TONIGHT. I'M FRANCOIS LAPIERRE, THE SPOKESPERSON FOR THE ASSOCIATION FOR ENVIRONMENTAL PROTECTION IN THE UPPER LAURENTIANS. IT'S MY PLEASURE TO INTRODUCE TO YOU OUR GUEST SPEAKER: GORDON EDWARDS FROM THE CANADIAN COALITION FOR NUCLEAR RESPONSIBILITY.
MONT-LAURIER, AUGUST 13TH 2006, QUÉBEC, CANADA
GOOD EVENING.TONIGHT I AM GOING TO TELL YOU ABOUT THE SECRET HISTORY OF CANADIAN URANIUM.
"IN AUGUST 1943 IN QUÉBEC CITY, CANADIAN PRIME MINISTER MACKENZIE KING, UK PRIME MINISTER WINSTON CHURCHILL AND US PRESIDENT F.D. ROOSELVELT SIGNED A SECRET AGREEMENT. IN IT, CANADA AGREED TO PROVIDE URANIUM TO WHAT BECAME KNOWN AS THE MANHATTAN PROJECT."
"IN TOTAL SECRECY, CANADIAN URANIUM WAS USED TO DEVELOP THE ATOMIC BOMBS DROPPED ON HIROSHIMA AND NAGASAKI ON AUGUST 6 AND 9, 1945, RESPECTIVELY."
"214,000 JAPANESE CIVILIANS DIED FROM THE BLAST AND EXPOSURE TO RADIATION."

"OLD MAN FERDINAND, VICTOR DOLPHU, JOE KENNY AND JOSEPH BAYHA, FOUR OF THE DENE WHO AS YOUNG MEN HAULED 100-POUND BAGS OF URANIUM ORE CONCENTRATE ALONG A 2,100-KILOMETRE TRANSPORTATION WEB OF RIVERS, RAPIDS AND PORTAGES KNOWN AS THE "HIGHWAY OF THE ATOM" FOR $3 A DAY. THEY ALL DIED OF CANCER."
'ON MAY 9, 2007, MINERS ARE STUNNED WHILE ENVIRONMENTALISTS CHEER OVER A NORTHERN REGULATOR'S RECOMMENDATION THAT A URANIUM EXPLORATION PROJECT BE DENIED BECAUSE IT THREATENS THE SPIRITUAL AND CULTURAL WELL-BEING OF THE AREA'S DENE PEOPLE.'
IN PORT RADIUM, A SHRINE IS DEDICATED TO THE MEMORY OF THE DEAD.
UNTIL HIS DEATH IN 1940, LOUIS AYAH, ONE OF THE NORTH'S GREAT ABORIGINAL SEERS, REPEATEDLY WARNED US THAT THE WATERS IN GREAT BEAR LAKE WOULD TURN A FOUL YELLOW. HE SAID THAT THERE WOULD BE SICKNESS AND OUR PEOPLE WOULD GO THROUGH HARD TIMES AND THAT THERE WOULD BE DEATHS.
MY HUSBAND, JOSEPH, WORKED FOR YEARS AT THE URANIUM MINE AND DIED AS MANY WHITE MINERS DID: COUGHING HIMSELF TO DEATH.
THE ELDORADO MINE WAS OWNED FIRST BY A PRIVATE COMPANY AND THEN BY A CROWN CORPORATION. IN 1988, THE OLD CROWN CORPORATION BECAME CAMECO LIMITED, THE WORLD'S BIGGEST URANIUM PRODUCER.
TODAY, CAMECO'S MARKET CAPITALIZATION REACHES $19.9-BILLION, WITH CORPORATE HEADQUARTERS IN SASKATOON, SASKATCHEWAN, THE U-CAPITAL OF THE WORLD.

IN MAY 2007, TORONTO GRAPHIC ARTIST RUTH TAIT AND MYSELF, SOPHIE TOUPIN, A JOURNALIST FROM MONTRÉAL, WENT TO MONT-LAURIER TO INTERVIEW KEY FIGURES RELATED TO THE URANIUM EXPLORATION IN THE AREA."
WE HAD 220 PEOPLE ATTENDING THIS PUBLIC INFORMATION SESSION LAST YEAR WITH GORDON EDWARDS.
MR. LAPIERRE, HOW DID YOU GET TO HEAR ABOUT NOVA-URANIUM'S EXPLORATION PROJECT IN MONT-LAURIER?
PEOPLE IN TOWN STARTED HEARING RUMORS ABOUT THE POSSIBILITY OF AN OPEN-PIT MINE. WE JUMPED ON THE INTERNET AND SAW THAT...
....A JUNIOR EXPLORATION COMPANY WAS EVALUATING THE LAND FOR URANIUM NEAR MONT-LAURIER.
30 YEARS AGO, I FOLLOWED THE SAME ISSUE. IT IS CYCLICAL. AT THAT TIME I THOUGHT URANIUM AND THE NUCLEAR INDUSTRY HAD COME TO A HALT, ESPECIALLY AFTER THREE MILE ISLAND AND CHERNOBYL.
BUT I GUESS THE NUCLEAR INDUSTRY IS DOING A PRETTY GOOD JOB TAKING ADVANTAGE OF THE KYOTO PROTOCOL AND THE SOARING PRICE OF URANIUM!.....
RIGHT NOW THERE ARE ABOUT 40 U-EXPLORATION PROJECTS THROUGH-OUT THE PROVINCE.....
.....BUT NO PRODUCTION.
TING! TIIING!

OUI, D'ACCORD.... AU REVOIR
ARE THERE ANY DANGERS TO COMMUNITIES LIKE MONT-LAURIER THAT ARE RELATED TO THE PROCESS OF URANIUM EXPLORATION?
LISTEN, RADON IS VERY DANGEROUS. WHEN THEY DRILL ON EXPLORATION SITES, PROSPECTORS ACTIVATE THE URANIUM, WHICH RELEASES TOXIC PARTICLES OF RADON INTO THE AIR AND INTO THE GROUND, POSSIBLY CONTAMINATING GROUNDWATER.
WHEN I SPOKE WITH LOCAL GEOLOGIST, ALAIN TREMBLAY, HE SAID THAT URANIUM EXPLORATION BENEFITS THE COMMUNITIES....
"ALAIN TREMBLAY: "IN THE REGION THEY ISSUE HOUSING CONSTRUCTION PERMITS WHILE KNOWING THERE IS A PROBLEM WITH RADON, A TOXIC GAS...."
EXPLORATION FOR URANIUM IS VERY USEFUL BECAUSE IT UNCOVERS RADON. WE DON'T WANT PEOPLE TO BUILD THEIR HOME ON URANIUM DEPOSITS THAT LEAK RADON.....

ALAIN TREMBLAY, DOWN-TOWN MT-LAURIER
THE SOCIAL RESPONSIBILITIES OF A JUNIOR COMPANY.....?
JUNIORS BOW DOWN TO THEIR SHAREHOLDERS....
SOCIALLY, JUNIORS HAVE AN INPUT FROM THE MOMENT THEY HIRE PEOPLE. THEY PAY LABOUR COSTS. THEY PAY THEIR TAXES...
....IN OTHER WORDS THEY ARE GOOD CORPORATE CITIZENS.
WHY SHOULD THEIR RESPONSIBILITY GO FURTHER THAN ANY OTHER CORPORATION?
WHEN THE PRICE OF URANIUM GOES UP IT'S BECAUSE THERE IS A NEED FOR MORE THAN THE MARKET CAN PROVIDE.
A JUNIOR COMPANY'S ROLE IS TO FIND MORE WHEN THE PRICE IS HIGH TO SUPPLY THE DEMAND AND TO MAKE PROFITS FOR THEIR SHAREHOLDERS... AND FUTURE PROFITS FOR URANIUM MINING COMPANIES.

"I SPOKE TO RICHARD DUFOUR, MANAGER AT NOVA-URANIUM. HE'S RESPONSIBLE FOR THE EXPLORATION PROJECT IN MONT-LAURIER."
THE REGION OF MONT-LAURIER IS GOING THROUGH A FOREST CRISIS. A U-EXPLORATION COMPANY SUCH AS NOVA-URANIUM IS BRINGING A LOT OF MONEY TO THE REGION.
IN 2006, NOVA SPENT ABOUT 5 MILLION DOLLARS IN MONT-LAURIER ON GAS, LODGING, AND ON EQUIPMENT RENTALS AND PURCHASES.
Mont-Laurier 123 Km
Mont-Tremblant Right
MONT-TREMBLANT IS GETTING ALL THE TOURISM THAT COMES OUT OF MONTRÉAL, NOT MONT-LAURIER.
"AND ANYWAY, QUEBEC IS A PRO-MINING PROVINCE. IT'S VERY FAVOURABLE FOR EXPLORATION COMPANIES. THE GOVERNMENT REIMBURSES 30% OF THE SPENDING DONE BY U-EXPLORATION COMPANIES AND UP TO 170% IS GIVEN IN TAX CREDITS TO INVESTORS WHO BUY MINING CERTIFICATES."
170%
TAX MINING
170% CREDIT
WHAT AM I GOING TO SAY TO THE MAYORS AT THE MAYORS' COUNCIL WHEN I MEET THEM?
I AM GOING TO GIVE THEM THE FACTS: ....THE APEHL ENVIRONMENTALISTS ARE DOING PROPAGANDA WORK!

DESPITE WHAT DUFOUR SAYS, LOCAL POLITICIANS HAVE COME ON BOARD TO HELP STOP THE PROJECT, LIKE MAYOR CLAUDE MENARD OF LAC ST-PAUL, MAYOR CLAUDE BLAIN OF LES CHUTES ST-PHILIPPE, AND SYLVAIN PAGÉ, THE PARTI QUÉBÉCOIS MP FOR ANTOINE-LABELLE AND A FOUNDING MEMBER OF APEHL.
LAST YEAR I ATTENDED AN INFORMATION SESSION ORGANIZED BY APEHL WHERE GORDON EDWARDS CAME TO TALK ABOUT URANIUM. THAT'S HOW I GOT TO KNOW ABOUT THE ISSUE.
SYLVAIN PAGÉ, MP
"IN NOVEMBER 2006, I PRESENTED A PETITION TO QUÉBEC'S NATIONAL ASSEMBLY, WHICH ORIGINATED FROM APEHL AND GATHERED 774 SIGNATURES. IT REQUESTED A MORATORIUM ON EXPLORATION, EXPLOITATION AND EXTRACTION OF URANIUM."
"I READ OUT THE PETITION AND NO QUESTIONS WERE ASKED. IT WAS TABLED, KEPT BY THE SECRETARIAT FOR 7 DAYS, THEN RETURNED TO ME WITHOUT A DEBATE. THE ACT OF PRESENTING A PETITION TO THE NATIONAL ASSEMBLY IS A LITTLE SAD."
I SENT LETTERS IN FEBRUARY AND MAY 2007 TO MINISTERS CLAUDE BÉCHARD, PHILIPPE COUILLARD, PIERRE CORBEIL AND LINE BEAUCHAMP TO INFORM THEM ABOUT THIS ISSUE.
I HAVE NOT YET RECEIVED AN ANSWER.
FOR THE MORATORIUM TO PASS, WE NEED STRONGER POPULAR SUPPORT. OUR EFFORTS SEEM TOO LOCAL. WE NEED TO GO PROVINCEWIDE.

"CLAUDE MÉNARD IS THE MAYOR OF LAC ST-PAUL WHERE THE U-EXPLORATION IS TAKING PLACE".
NOVA IS DOING EXPLORATION WORK IN MY MUNICIPALITY. I HAVE PRESENTED A PLEA TO THE MUNICIPALITY FOR A MORATORIUM IN LAC ST-PAUL.
"CLAUDE BLAIN IS MAYOR OF LES CHUTES ST-PHILIPPE..."
AS AN ADJOINING MUNICIPALITY, WE SUPPORT MR. MÉNARD'S REQUEST FOR A MORATORIUM. THERE ARE MANY REASONS TO BE OPPOSED TO THIS EXPLORATION PROJECT, ESPECIALLY IF IT LEADS TO AN OPEN-PIT MINE.
I HAVE SEEN THE DANGERS WITH ELEMENT 92.
BEFORE RETIRING IN THE REGION, I WORKED FOR A LONG TIME AT THE CHALK RIVER POWER STATION.
THERE, I SAW A MAN DIE IN THREE DAYS FROM EXPOSURE TO RADIATION.

"TO GET ANOTHER PERSPECTIVE, I PHONED JAMES FINCH, AN AUTHOR AND URANIUM ANALYST."
Platinum Author Expert
QUEBEC HAS ONE OF THE MOST PROLIFIC MINING HISTORIES IN THE WORLD. MY GOD! LOOK HOW MUCH MINERAL RESOURCES IT SUPPLIES, IT'S AMAZING, AND IT'S A GREAT PLACE FOR THAT. AS FOR U-MINING I DON'T KNOW. IT'S SO LOW GRADE....
WITH THE PRICE OF URANIUM SOARING FROM $20 A POUND IN 2005 TO $138 IN 2007, IT SEEMS MORE THAN WORTH IT FOR A JUNIOR COMPANY TO DO EXPLORATION WORK, NO?
....WELL SOPHIE,... THEY MIGHT GIVE YOU THAT IMPRESSION, BUT GET A COUPLE OF THEM DRUNK AND YOU'LL GET A DIFFERENT STORY!
CLIK! CLIK!... CLIK!... CLIK!
"JUNIOR URANIUM EXPLORATION COMPANIES, FOR THE MOST PART, ARE CREATED FOR THE SOLE PURPOSE OF GIVING JOBS TO UNEMPLOYED PEOPLE. THERE ARE SO FEW LEGITIMATE COMPANIES OUT THERE."
ONCE YOU START DISCOVERING URANIUM AND ACTUALLY GO INTO PRODUCTION; THAT CAN TAKE 10 TO 20 YEARS.
MOST OF THESE GUYS WILL BE DEAD BY THEN!
TAP! TAP! TAP! TAP!
"AS I REVIEWED MY CONVERSATION WITH JAMES FINCH, I WAS REMINDED OF AN ARTICLE IN THE JOURNAL LES AFFAIRES THAT READ..."
LES AFFAIRES
April 22, 2006
Uranium Prospecting in Québec is on The Rise.
Nova Uranium evaluates the possibility of an open-pit near Mont-Laurier
As a consequence of this price increase, a dozen new exploration projects are underway in Québec, according to Mr. Thomassin, Executive Director of The Quebec Mineral Exploration Association...

URANIUM IN MONT-LAURIER WAS DISCOVERED IN 1967. HOWEVER, PRIOR TO 1973, URANIUM WAS CONTROLLED BY THE MILITARY AND A MARKET DID NOT YET EXIST.
BACK TO GORDON EDWARDS.
"THE PETROLEUM CRISIS OF 1973 CAUSED URANIUM PRICES TO SURGE FROM US$6 TO US$45 A POUND. BUT IN 1978, THE ACCIDENT AT THREE MILE ISLAND LED ALMOST TO A FULL CESSATION OF ACTIVITIES.
THREE MILE ISLAND
ACTION NOW!
NO NUCLEAR
BETWEEN 1970 AND 1984, U-PRODUCTION EXCEEDED THE DEMAND. THIS WAS BASED ON THE ERRONE-OUS PREDICTIONS OF THE GROWTH OF THE NUCLEAR ENERGY SECTOR. FROM 1985 TO 2003,
LIQUIDATION OF INVENTORIES DOMINATED THE SCENE.
SINCE 2004, SEVERAL NEW MINING CLAIMS, ESPECIALLY IN THE PROVINCE OF QUEBEC, WERE ACQUIRED.
IN THE MONT-LAURIER REGION, NOVA-URANIUM CORPORATION, STRATECO RESOURCES INC., CONSOLIDATED PACIFIC BAY MINERALS LTD. AND BIG RED DIAMOND CORPORATION, AMONG OTHERS, HAVE CLAIMS FOR U-EXPLORATION.

QUEBEC HAS A VERY GOOD JURISDICTION FOR MINING, INCLUDING ENCOURAGEMENT IN THE FORM OF REFUNDS FOR EXPLORATION EXPENSES AND TAX INCENTIVES FOR DEVELOPMENT.
BACK TO JAMES FINCH
AND JUST LOOK HOW CLOSE ONTARIO IS! IT'LL BE A NATURAL MARKET IF A MINE IS DEVELOPED IN MONT-LAURIER.
FINCH'S COMMENTS REMINDED ME THAT NUCLEAR ENERGY ALREADY PROVIDES ABOUT 15% OF CANADA'S ELECTRICITY, AND 51% OF ONTARIO'S ENERGY CONSUMPTION COMES FROM NUCLEAR POWER.
U-COMPANIES WORK ON LONG-TERM CONTRACTS WHERE MOST OF THE URANIUM TRADED IS CONTRACTED WELL IN ADVANCE. THE SPOT PRICE IS NOW AT $138 WHEREAS THE AVERAGE PRICE THE MAJOR PRODUCERS ARE GETTING ARE IN THE TWENTIES!
"$20/POUND...DELIVERY 2008.... AGREED...."
THE SPOT PRICE IS BASICALLY FOR THE END USERS OF URANIUM THAT DON'T HAVE ALL THE URANIUM THEY NEED. EITHER THEY WERE NOT CLEVER ENOUGH. OR,
THEY WERE SIMPLY NOT WELL PREPARED. THEY DRIVE THE SPOT PRICE HIGHER BECAUSE THEY NEED TO GET SOME NOW!
THEY NEED SOME FOR THEIR COMMITMENTS, LET'S SAY 2 TO 3 YEARS DOWN THE ROAD, WHICH IS THE TIME IT TAKES AFTER EXTRACTION TO ENRICH THE URANIUM BEFORE IT ACTUALLY GETS INTO A REACTOR.
THERE IS A LOT MORE VOLUME IN OTHER COMMODITY MARKETS—LIKE COPPER—THAN IN THE URANIUM MARKET BECAUSE COPPER IS USED FOR EVERYTHING.
URANIUM IS USED IN TWO THINGS....
....NUCLEAR BOMBS AND NUCLEAR REACTORS.

"APEHL HAS DISCUSSIONS WITH OTHER COMMUNITY GROUPS TO DEVELOP A COALITION AGAINST THE URANIUM MINING PROJECT. DEMOCRATIC PROCESSES HAVE MADE NO DIFFERENCE DESPITE SUPPORT FROM LOCAL/PROVINCIAL POLITICIANS. THE U-MINING PROJECT SEEMS TO BE MOVING FORWARD..."
AT THE MAY 31, 2007 APEHL MEETING, ONLY FIVE PEOPLE SHOWED UP.
MR. RICHER, YOU HAD CONTACTS WITH AN INDIVIDUAL LOCATED NEAR OTTAWA THAT IS RESISTING A URANIUM EXPLORATION PROJECT. CAN YOU TELL US A LITTLE BIT MORE ABOUT IT....
I CONTACTED MICHAEL PATENAUDE, A CITIZEN FROM THE VILLAGE OF LA PÊCHE, WHO WAS TOLD THAT THE LAND ON WHICH HE HAD HIS COTTAGE WAS CLAIMED FOR MINING ACTIVITIES.
MICHAEL WAS TELLING ME THAT ACCORDING TO ARTICLE 235 OF THE QUÉBEC MINING ACT: THE HOLDER OF MINING RIGHTS MAY ACQUIRE BY AGREEMENT OR EXPROPRIATION ANY PROPERTY PERMITTING ACCESS FOR EXPLORATION WORK OR MINING OPERATIONS.
THIS MEANS THE SURFACE OF THE LAND BELONGS TO THE OWNER OF THE PROPERTY, BUT THE MINERAL RIGHTS BENEATH THE SURFACE DON'T. AND SO ANYONE COULD CLAIM IT, IF THEY SUCCESSFULLY APPLIED FOR A PROSPECTOR'S LICENSE. AND MIND YOU, NONE OF THE PROFIT WOULD BE SHARED WITH THE PROPERTY OWNERS!
"IN THE AYLMER SECTOR OF THE CITY OF GATINEAU, ECO-WATCH IS LOBBYING TO GET THE QUÉBEC MINING ACT AMENDED, TO DIMINISH THE POWER OF MINING RIGHTS HOLDERS."

AT ANOTHER APEHL MEETING ON JUNE 26, 2007....
WE URGENTLY NEED TO CONVINCE MORE MAYORS TO SUPPORT A MORATORIUM AND CREATE A COMMUNICATION COMMITTEE THAT WILL BE MANDATED TO DRAFT A STRATEGY.
WE HAVE TO DEVELOP MORE LINKS WITH GROUPS FIGHTING OTHER U-PROJECTS SUCH AS MICHAEL IN OTTAWA OR THE PEOPLE FROM KITIGAN ZIBI RESERVE NEAR MANIWAKI.
MAY 2007, FRANÇOIS LAPIERRE PRESENTS APEHL'S CASE ON CKUT 90.3 FM AND RADIO CENTRE-VILLE, TWO INDEPENDANT RADIO STATIONS IN MONTRÉAL.
WE NEED TO REACH OUT TO THE PEOPLE OF MONTRÉAL, QUÉBEC CITY AND THE REST OF THE PROVINCE. WE MUST HARNESS THE POWER OF THE MEDIA TO GET OUR STORY OUT.
APEHL LOBBIED AT MUNICIPAL AND PROVINCIAL LEVELS BUT NO ONE SEEMS TO BE LISTENING. WHAT DO COMMUNITIES NEED TO DO TO GET HEARD WHEN POLITICIANS IN POSITIONS OF POWER ARE NOT INTERESTED IN THEIR PERSPECTIVE?
....I READ IN THE NEW INTERNATIONALIST THAT "DURING THE PERIOD OF 1953-2002, THE GOVERNMENT OF CANADA, WITH THE TAXPAYERS' MONEY, GAVE THE NUCLEAR INDUSTRY AROUND 14.5 BILLION DOLLARS."
WHAT? PRIVATIZING THE CANADIAN ATOMIC ENERGY AGENCY? AFTER ALL THESE FUNDS HAVE BEEN POURED INTO RESEARCH AND IMPLEMENTATION? NO, I HAVEN'T SEEN THE TORONTO STAR ARTICLE. JULY 6? OK, I'LL CHECK IT OUT.
NOW WITH CO2 EMISSIONS BEING THE LATEST ENVIRONMENTAL BAD GUY, NUCLEAR POWER IS BEING PROMOTED AS THE "GREEN" ENERGY. REMEMBER WHAT DALTON MCGUINTY, PREMIER OF ONTARIO SAID IN FEBRUARY 2003: "NUCLEAR POWER STATIONS GENERATE CLEAN ELECTRICITY....
...COMPARED TO COAL AND OIL, THUS BRINGING DOWN GREENHOUSE GAS EMISSIONS AND RESPECTING KYOTO'S PROTOCOL." JUST WATCH AS WORD GETS OUT ABOUT NUCLEAR REACTORS BEING BUILT TO SUPPLY ENERGY FOR TAR SANDS EXTRACTION IN ALBERTA.

"URANIUM TRADING IS A SECRET MARKET. IT'S THE MOST SECRET BECAUSE IT'S THE MOST REGULATED MARKET."
"YOU AND I CAN'T GO AND START MINING URANIUM, YOU NEED LICENCES ALL OVER THE PLACE."
IN TERMS OF THE MARKET, ACTIVITIES ARE BOOMING. THERE ARE LOADS OF CONSOLIDATIONS IN THE U-INDUSTRY.
JAMES FINCH
"NOW THAT THE TORONTO STOCK EXCHANGE HAS JUST LAUNCHED A GLOBAL MINING INDEX, IT WILL DRAW INVESTMENT FROM COMMODITY-HUNGRY STOCK BUYERS."
IT'S GREAT NEWS!
IF I HAD TO DIVERSIFY MY PORTFOLIO, I WOULD BUY STOCKS FROM HUGE TRANSNATIONAL CORPORATIONS SUCH AS GENERAL ELECTRIC AND TOSHIBA AS THEY HAVE NEW PRODUCTS ON THE MARKET:
NUCLEAR REACTORS.
I WILL CONCLUDE THE EVENING QUOTING THE HARVARD BUSINESS SCHOOL TASK FORCE REPORT ON ENERGY 2006 WHICH HAS ARGUED THAT "WE CAN LIVE QUITE AFFLUENTLY WITHOUT REQUIRING MORE NUCLEAR POWER, OIL OR COAL, BY INVESTING IN ENERGY EFFICIENCY, ENERGY CONSERVATION, AND RENEWABLE FORMS OF ENERGY."
THANK YOU.

BAUXITE

BAUXITE

The World's Unluckiest People

REPORTAGE BY TAMARA HERMAN
COMIX BY STANLEY WANY

BAUXITE

This is the story of Kashipur, India. It is a long story, filled with victories and suffering, repression and uprising. It is a story about power and survival, about oppression and resistance. It is a story that ties continents together. Visitors to Kashipur may leave overcome with sadness or overjoyed with hope. They will not be left untouched. The story began before this journalist came to Kashipur, and continues long after she left. But we must start somewhere, and January 2005 is our beginning, in India, on a side-road between Rayagada and Barigan.
On a typical 36° day in Kashipur, our guide looks out towards the peak of Baphlimali Mountain.

The people here are the unluckiest people in the world.
Unlucky? This looks like paradise!
It's because their lives and livelihoods are threatened. The bauxite reserves on which they live have been a point of contention for many years. These villagers are facing large companies that want to build a mine and smelter.
The government supports the companies, and not the villagers. And the government has police.
The Canadian company says that the villagers consent to the mine, but other reports say that the locals refuse to accept compensation or resettlement. Which of these stories is true?
The people will tell you themselves. As you will see, we are surrounded by police, by the Indian Reserve Battalion, by the company's hired goons. People have died protesting the project. Others have been injured, have gone to jail, or have seen their villages torn apart and divided. The state uses all its forces to suppress the people.
But still, the people will not be silenced.

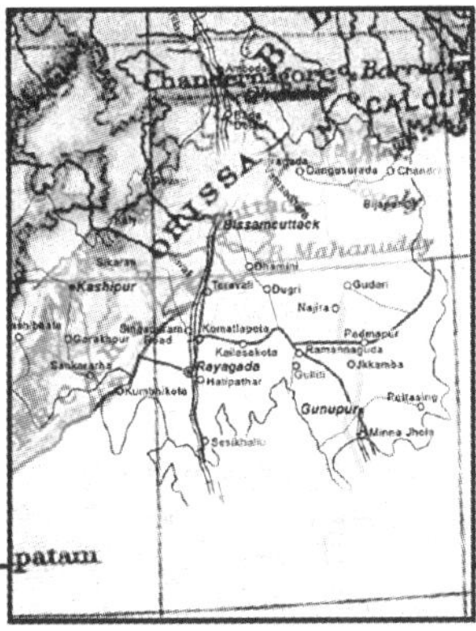

In 1993, a conglomerate of corporations created Utkal Alumina International Limited (UAIL) to mine Baphlimali Mountain of its bauxite. As part of the project, alumina (the raw material for aluminum) is to be extracted from the bauxite at a refinery proposed to be built near the mine site in Kashipur. UAIL was initially a joint venture between India's TATA, Norway's Norsk Hydro, India's Indal. Canada's Alcan joined the partnership in 1998.The mountain was calculated to hold 70% of India's bauxite reserves and 14% of world's total deposits. It is estimated that the project could process one million tonnes of bauxite per year, generating up to one billion dollars. The immense project is supported by both the government of the State of Orissa and the local government.

Baphlimali is a sacred hill. Our ancestors won this hill from the clutches of the foreign colonials when they were at work in our area. Our forefathers fought for that hill and it is our responsibility to ensure that our mountain does not go into the hands of others.

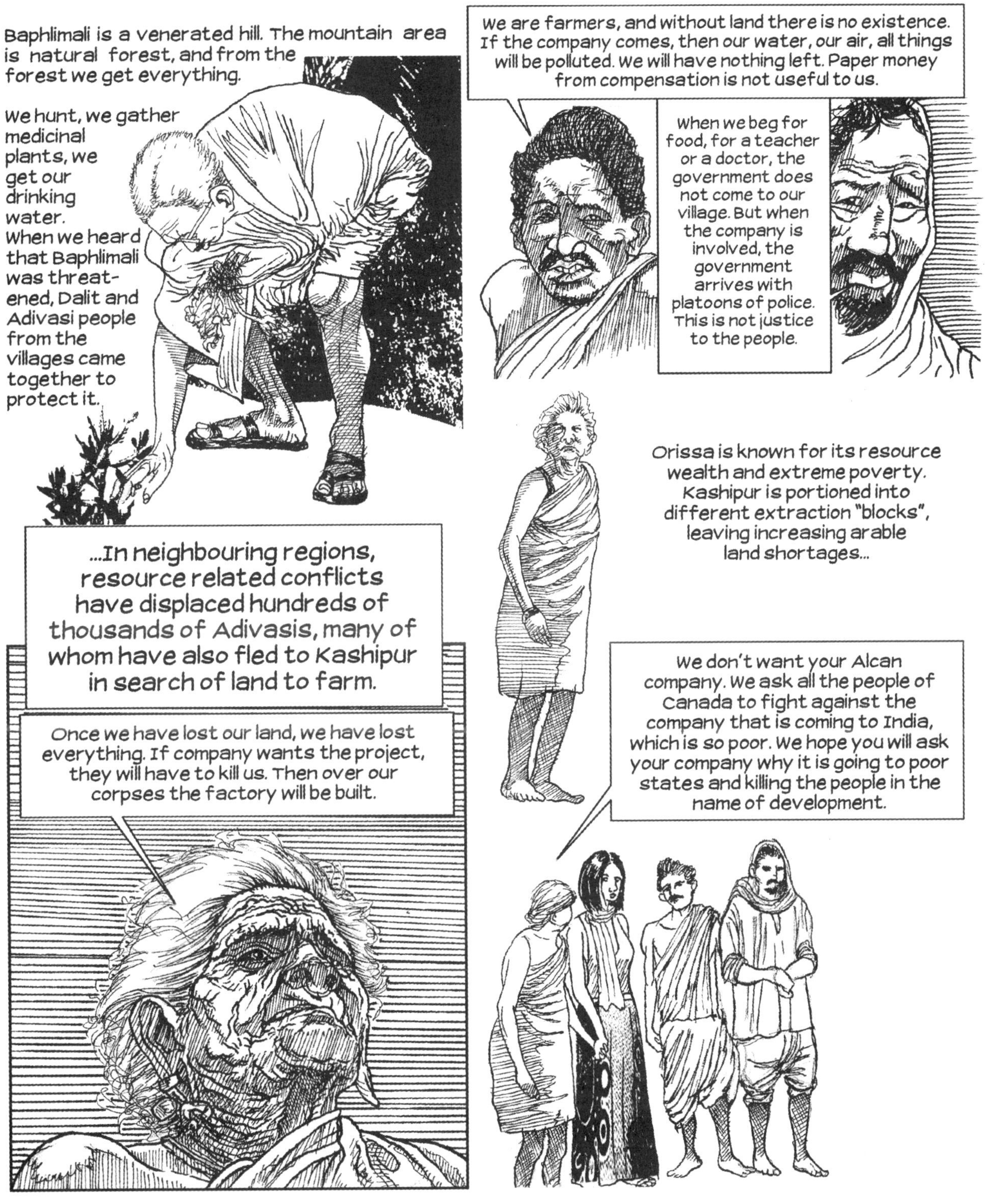
Baphlimali is a venerated hill. The mountain area is natural forest, and from the forest we get everything.
We hunt, we gather medicinal plants, we get our drinking water. When we heard that Baphlimali was threatened, Dalit and Adivasi people from the villages came together to protect it.
We are farmers, and without land there is no existence. If the company comes, then our water, our air, all things will be polluted. We will have nothing left. Paper money from compensation is not useful to us.
When we beg for food, for a teacher or a doctor, the government does not come to our village. But when the company is involved, the government arrives with platoons of police. This is not justice to the people.
Orissa is known for its resource wealth and extreme poverty. Kashipur is portioned into different extraction "blocks", leaving increasing arable land shortages...
...In neighbouring regions, resource related conflicts have displaced hundreds of thousands of Adivasis, many of whom have also fled to Kashipur in search of land to farm.
Once we have lost our land, we have lost everything. If company wants the project, they will have to kill us. Then over our corpses the factory will be built.
We don't want your Alcan company. We ask all the people of Canada to fight against the company that is coming to India, which is so poor. We hope you will ask your company why it is going to poor states and killing the people in the name of development.

Look out the window. See how arid it is here? There has been starvation, famine in Kashipur. Here, communities rely on subsistence agriculture, hunting and gathering. When people's survival is tied to the land in a fragile ecosystem, the water shortages and contamination, soil contamation and deforestation that may come from the mine are very dangerous. Studies show that the watershed is linked to Baphlimali. Do you see why it is a matter of survival?
Could this be why Alcan didn't want to release the environmental impact assessment?
Professor Rath is a retired lecturer and a solidarity activist in the struggles of Orissa's Adivasi and Dalit communities fighting resource extraction. He now lives in a small house in Rayagada, the largest town closest to Kashipur.
In India, it has been estimated that 50 million people have been forcibly displaced for 'development' projects, and almost 3 million for mines. Many of these people are now residing in crowded city slums and suffering terribly.
In the beginning, people in Kashipur were in favour of mining because they thought it would bring development.
But as days passed people became more conscious. And when they saw others suffering in other parts of Orissa, they decided that they would fight.
These mines will devastate the lives of people. Their lives would be totally destroyed. And that is why they are fighting.

In Rayagada, I am staying with a family that runs a non-governmental organization that works with Adivasi communities working to gain access and control over political and economic resources.
Even though Orissa is rich in resources, its people are poor. We Adivasi have seen that wherever big companies come to work, there are many problems. Across Orissa, tribal peoples have lost their land, homes, culture, existence.
At the forefront of the battle for bauxite is the thorny matter of community consent. Several Indian laws give special protection to Adivasi land. The Fifth Schedule of the Constitution holds that official local consent must be given for resource extraction.
Open meetings in which all villagers participate freely must be held to issue this consent. But consent can be a difficult matter. When it can not be easily found, some may tell you that it can be fabricated.

The next day, we set out again to visit affected villages. We pull into Maikanch shortly after police have left. They had oc-cupied the village for almost two months. A Dalit villager speaks to us not far from a monument erected to commemorate the death of three Adivasi men from Maikanch. They were shot by police in December 2000, while protesting the mining project.
In the years before our brothers were killed, tension was building in Kashipur. The Prakrutik Sampad Suraksha Parishad, or the PSSP, was among the groups formed to fight the mining project. We had many tactics, from negotiations, to rallies, to roadblocks.

But the company was also at work, and I am convinced that it asked the government to send the police. The police terrorized our villages, arresting people, bribing people, creating factions and divisions. The militarization of the region had begun.
As trouble grew throughout the region, the company tried to stage a pro-mine meeting with important officials near Maikanch. The Dalit and Adivasi made a roadblock, and we stopped the meeting. The company and the government were very angry. It was December 15, 2000.
They sent the police the next day. December 16th 2000 was the day of the shooting, the day of terror.
BANG!!
BANG!!

Their names were Abhilash Jhodia, Raghunath Jhodia and Damodar Jhodia.
They are the martyrs who were shot by police. We honour them every year...
Many others were injured in the Maikanch firings. Arjun is my son, and he can no longer work because the police fired a bullet into his leg. We are waiting for compensation.
You see this hole in my leg? Also from a bullet.
The Maikanch killings drew attention to the crisis in Kashipur, and solidarity campaigns were launched across the globe. A group named "Alcan't in India" was formed in Montréal, where Alcan has its head-quarters. The group held protests and pressured the corporation to divest.
A shareholder's activist group bought Alcan shares to exert pressure from the inside.
NO ALCAN ON OUR LAND
ALCAN Va t'en!
In Norway, a group called NorWatch pressured Norsk Hydro to withdraw from the project. When both Norsk Hydro and India's TATA had indicated their intent to withdrew in 2001, Alcan's stake in UAIL was increased to 45%.The UAIL project was put on hold until the judicial inquiry into the killings was completed.
THE HINDU
INDIA'S NATIONAL NEWSPAPER
ALCAN GOES GREEN
By Our Staff Reporter
The Gazette
MONTREAL | MONDAY, JUNE 12, 2006 | montrealgazette.com | SINCE 1778 | SPORTS FIN
ALCAN GOES GREEN
Alcan said that its final approval would depend on the results of the inquiry. In the meantime, it claimed to be committed to economic, environmental and social sustainability. But the uneasy calm that had settled over Kashipur dissipated when the inquiry put its criticisms aside to give UAIL the go-ahead.
Indian media reported Orissa Chief Minister Naveen Patnaik saying,
"Anybody coming in the way of mining and industrialization will be severely dealt with."

We return to Rayagada, with some answers and many more questions. We talk with Sibaram, a villager and anti-mine activist from Maikanch.
The Maikanch inquiry was held and the project was allowed to proceed. Trouble started soon af-terwards. UAIL has not yet gotten the consent of the people, which is mandatory. So the company felt pressured and held its consultation meeting.
At consultation meetings, known as Gram Sabhas, Adivasi and Dalit must give their free consent for any development on their land. But this particular meeting was a stage for corruption, harassment and coercion. Not all families were invited and the Gram Sabha was filled with police. It was called by politicians, not villagers. Some people said they were forced to sign their names.
And then, weeks later, the company orchestrated a pro-mine rally. It paid people to come and show their support. Witnesses say that the people were from outside the UAIL block.
Can you prove this?
We don't have proof of whether a person has received 50 Rupees or 100 Rupees to come to the rally. There is seldom easy proof of bribery. But we have proof that dozens of trucks were hired. All of Orissa saw them on television. At 'real' gram sabhas, people walk. Why? Because trucks do not come for free.

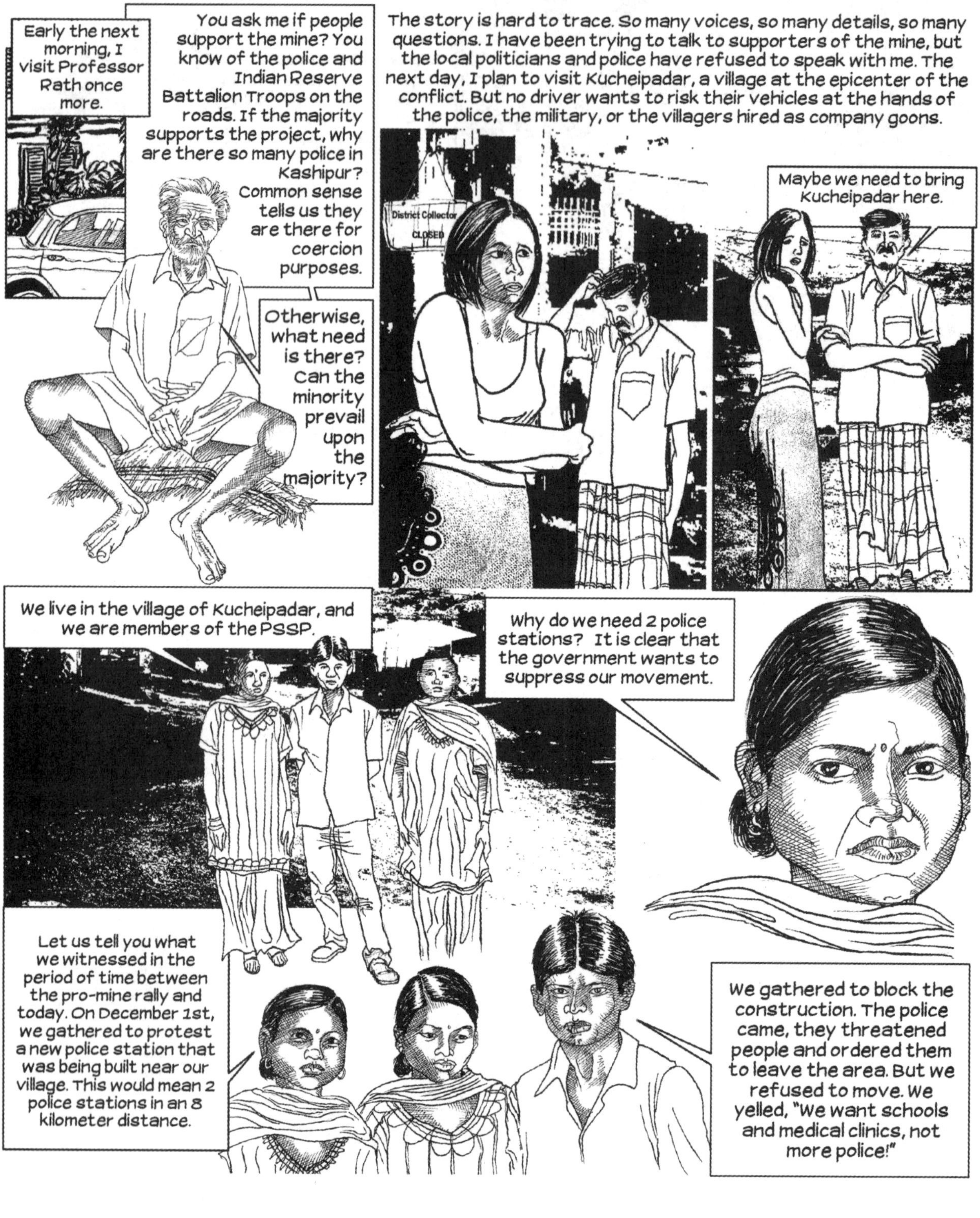
Early the next morning, I visit Professor Rath once more.
You ask me if people support the mine? You know of the police and Indian Reserve Battalion Troops on the roads. If the majority supports the project, why are there so many police in Kashipur? Common sense tells us they are there for coercion purposes.
Otherwise, what need is there? Can the minority prevail upon the majority?
The story is hard to trace. So many voices, so many details, so many questions. I have been trying to talk to supporters of the mine, but the local politicians and police have refused to speak with me. The next day, I plan to visit Kucheipadar, a village at the epicenter of the conflict. But no driver wants to risk their vehicles at the hands of the police, the military, or the villagers hired as company goons.
District Collector
CLOSED
Maybe we need to bring Kucheipadar here.
We live in the village of Kucheipadar, and we are members of the PSSP.
Why do we need 2 police stations? It is clear that the government wants to suppress our movement.
Let us tell you what we witnessed in the period of time between the pro-mine rally and today. On December 1st, we gathered to protest a new police station that was being built near our village. This would mean 2 police stations in an 8 kilometer distance.
We gathered to block the construction. The police came, they threatened people and ordered them to leave the area. But we refused to move. We yelled, "We want schools and medical clinics, not more police!"

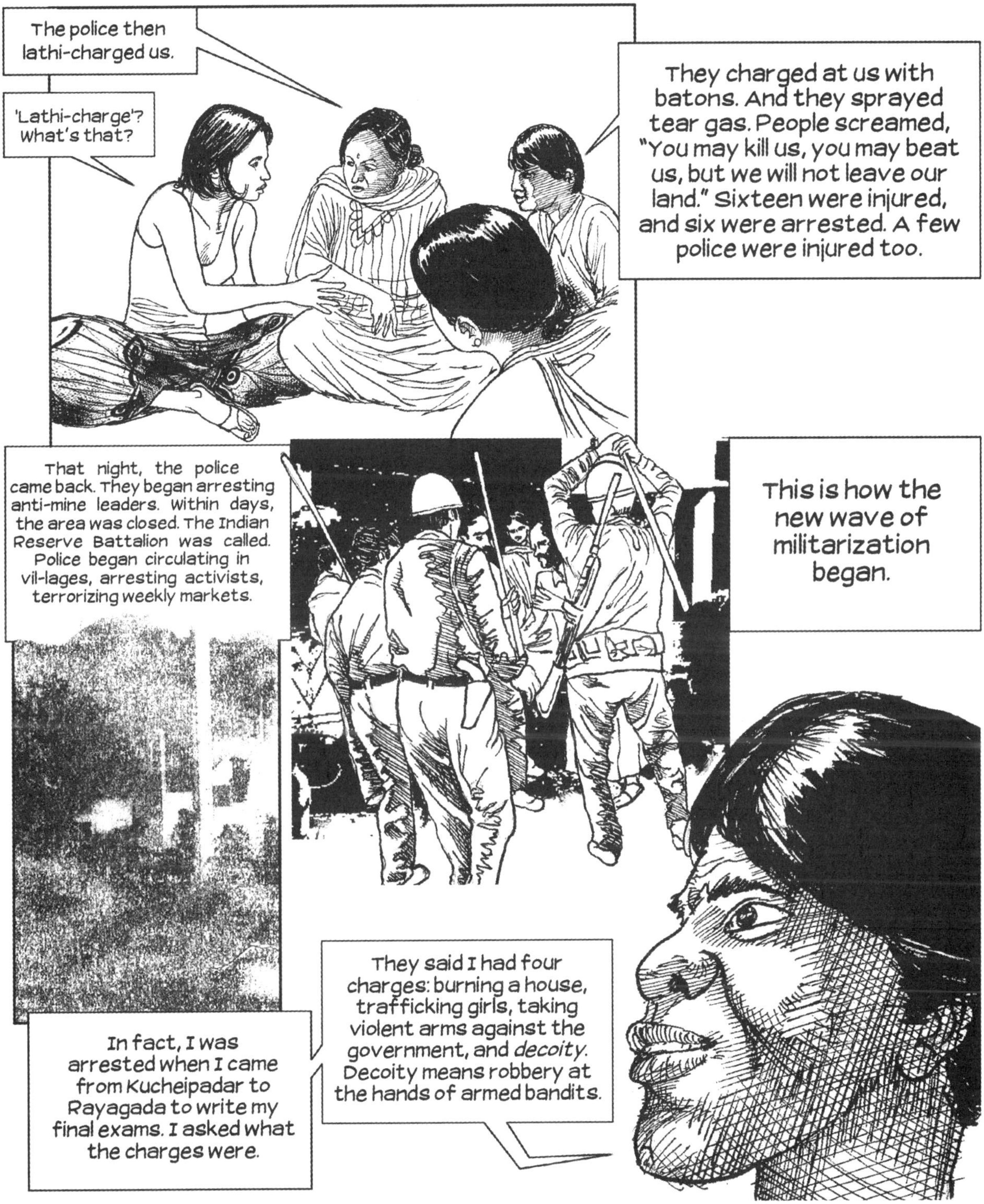

The police then lathi-charged us.
'Lathi-charge'? What's that?
They charged at us with batons. And they sprayed tear gas. People screamed, "You may kill us, you may beat us, but we will not leave our land." Sixteen were injured, and six were arrested. A few police were injured too.
That night, the police came back. They began arresting anti-mine leaders. Within days, the area was closed. The Indian Reserve Battalion was called. Police began circulating in vil-lages, arresting activists, terrorizing weekly markets.
This is how the new wave of militarization began.
In fact, I was arrested when I came from Kucheipadar to Rayagada to write my final exams. I asked what the charges were.
They said I had four charges: burning a house, trafficking girls, taking violent arms against the government, and *decoity*. Decoity means robbery at the hands of armed bandits.

The government cannot buy our people. It can hire people from outside to come to a rally and pretend that people are welcoming the project. The government can do fraud, it can try to tear our communities apart with bribery, repression, and harassment. But it cannot buy our people.
And so we challenge the government to let the people from our area express our consent or dissent freely, without bribery, corruption nor intimidation. The government may attempt to silence us, but our movement continues.
We have had a freedom struggle for over 200 years. It is very meaningful for our existence. Should our struggle end, we would not yet have fully realized our freedom.

Naveen is a lawyer working with three of the thirteen anti-mine villagers held in jail. He takes me to the jail to visit the prisoners, including 74-year-old Adivasi leader Bulka Miniaka of Barigan. We are not allowed in the cells, but wardens bring several prisoners to us.

We don't want to welcome a company that does not respect our life and property. We only want a peaceful life.
Will the people's movement stand up to this pressure and coercion?
It is difficult to say. The government may shoot at people, it may confine people, it may send more villagers to jail. In this way, they may succeed, or they may not succeed. The state seeks the advancement of capitalism and globalization, and to suppress the Kashipur struggle they are using all their forces: judiciary, security, political parties...
It is the tribal communities fighting Alcan and Hindalco. It is good they have continued their fight. Because of the community's efforts the company has still not started work.
We rely very much on solidarity. If Alcan withdraws, it is an achievement, but another company will come. We must then oppose other multinationals, and not just Alcan. And we must have links to other communities who share our struggles.
If we continue this process, one day we will have very much strength.

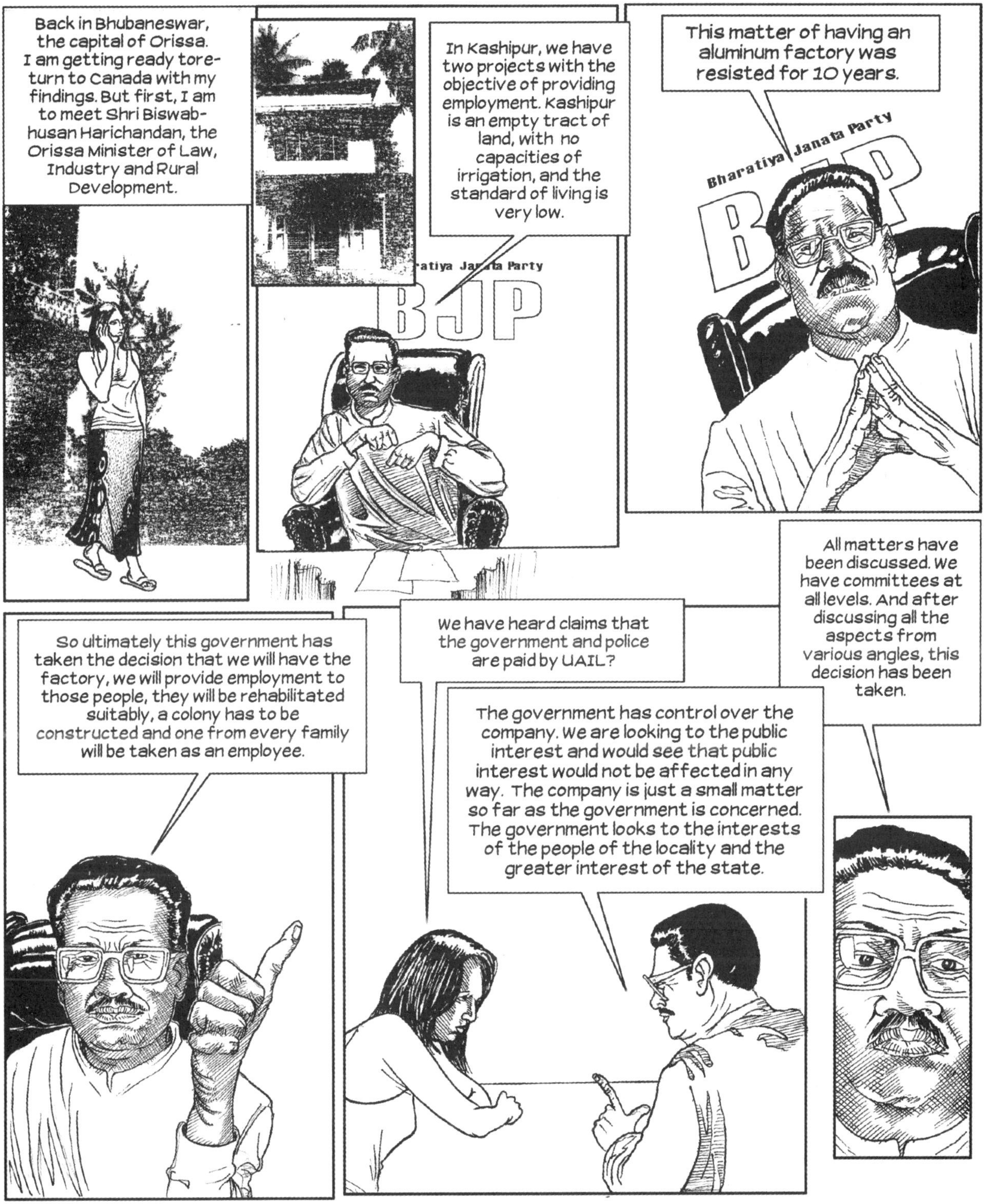

Back in Bhubaneswar, the capital of Orissa. I am getting ready tore-turn to Canada with my findings. But first, I am to meet Shri Biswab-husan Harichandan, the Orissa Minister of Law, Industry and Rural Development.
In Kashipur, we have two projects with the objective of providing employment. Kashipur is an empty tract of land, with no capacities of irrigation, and the standard of living is very low.
Bharatiya Janata Party
BJP
This matter of having an aluminum factory was resisted for 10 years.
Bharatiya Janata Party
All matters have been discussed. We have committees at all levels. And after discussing all the aspects from various angles, this decision has been taken.
So ultimately this government has taken the decision that we will have the factory, we will provide employment to those people, they will be rehabilitated suitably, a colony has to be constructed and one from every family will be taken as an employee.
We have heard claims that the government and police are paid by UAIL?
The government has control over the company. We are looking to the public interest and would see that public interest would not be affected in any way. The company is just a small matter so far as the government is concerned. The government looks to the interests of the people of the locality and the greater interest of the state.

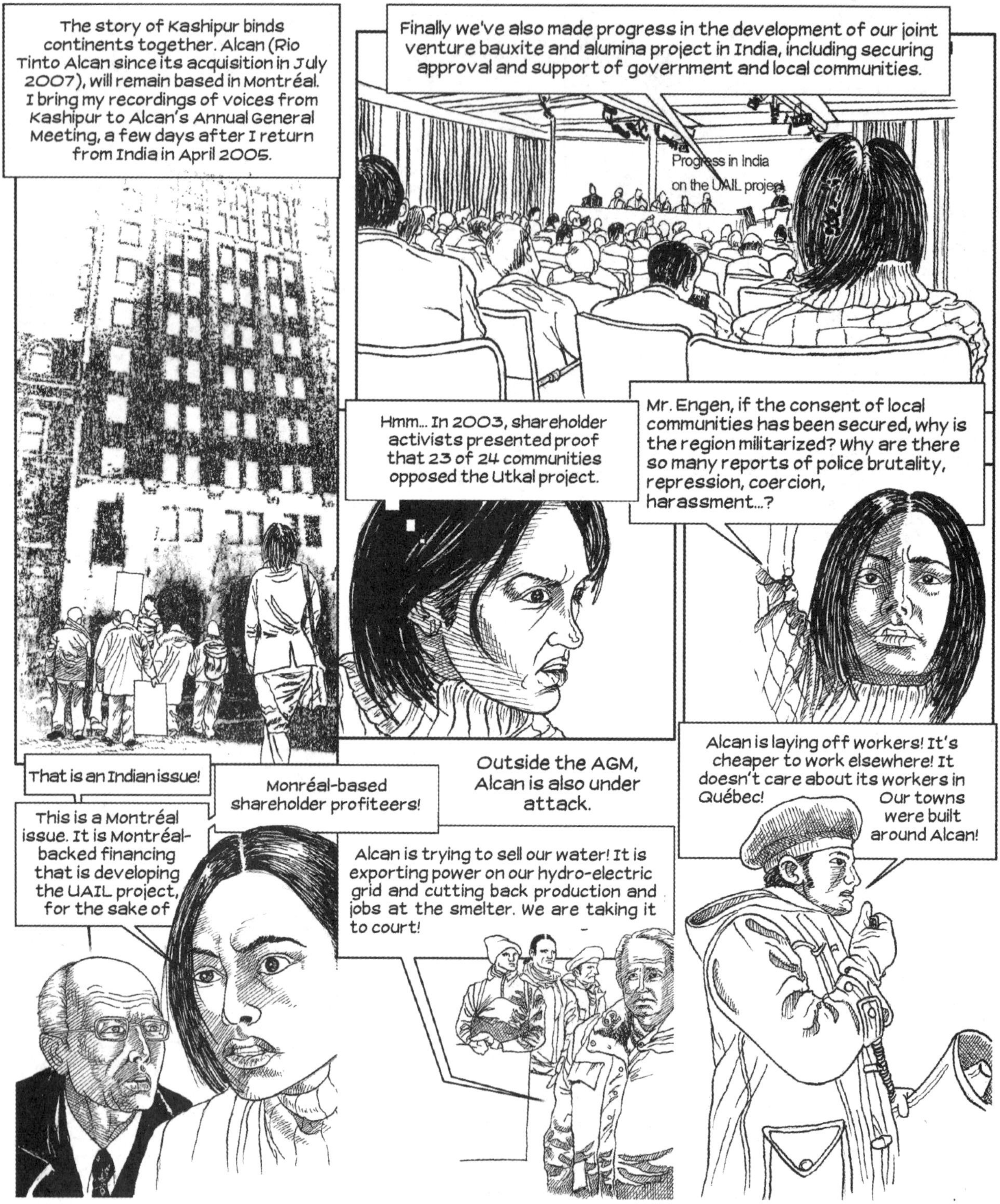

The story of Kashipur binds continents together. Alcan (Rio Tinto Alcan since its acquisition in July 2007), will remain based in Montréal. I bring my recordings of voices from Kashipur to Alcan's Annual General Meeting, a few days after I return from India in April 2005.
Finally we've also made progress in the development of our joint venture bauxite and alumina project in India, including securing approval and support of government and local communities.
Progress in India
on the UAIL project
Hmm… In 2003, shareholder activists presented proof that 23 of 24 communities opposed the Utkal project.
Mr. Engen, if the consent of local communities has been secured, why is the region militarized? Why are there so many reports of police brutality, repression, coercion, harassment…?
That is an Indian issue!
Monréal-based shareholder profiteers!
This is a Montréal issue. It is Montréal-backed financing that is developing the UAIL project, for the sake of
Outside the AGM, Alcan is also under attack.
Alcan is trying to sell our water! It is exporting power on our hydro-electric grid and cutting back production and jobs at the smelter. We are taking it to court!
Alcan is laying off workers! It's cheaper to work elsewhere! It doesn't care about its workers in Québec!
Our towns were built around Alcan!

August, 2005
Dear friends in India,
The battlefield for bauxite is a public relations war here. Alcan is trying desperately to come across as a socially responsible corporation. It is even offering an Alcan Prize for Sustainability!
August 2005...
Dear friends in India,
The battlefield for bauxite is a public relations war here. Alcan is trying desperately to come across as a socially responsible corporation. It is even offering an 'Alcan Prize for Sustainability'!
Alcan sustainability report
Workers from Alcan plants in Québec and British Columbia have passed resolutions to refuse to smelt any alumina shipped from Kashipur! Resistance against Alcan is moving up the production cycle. Not only is there resistance in India, there is also resistance in Canada at major Alcan smelters! Next, we need to target the manufacturers! And keep pressure on consumers, which are all of us...
Bauxite mining
alumina production
primary aluminum production
recycling
use phase
semi-fabrication
product manufacture
April, 2006
Dear friends in Canada,
I am delighted to send you this report from the Orissa Legislative Assembly. It criticizes the project on social and environmental grounds, and confirms that UAIL's environmental clearance has expired. This means the infrastructure construction now underway is illegal!
Violence has broken out in nearby regions, and the communities continue to live in terror. But you will be happy to know that no fewer than five fact-finding missions conducted by respectable organizations have denounced the use of repression and police brutality to obtain the consent of affected villagers in Kashipur.
April 2006.
The campaign in Canada against Alcan continues, inside and outside the corporation's walls.
As you are aware, we presented a shareholder's proposal demanding an independent advisory committee to study the project. Our proposal gained 36% of the votes. This is an exceptional level of support!

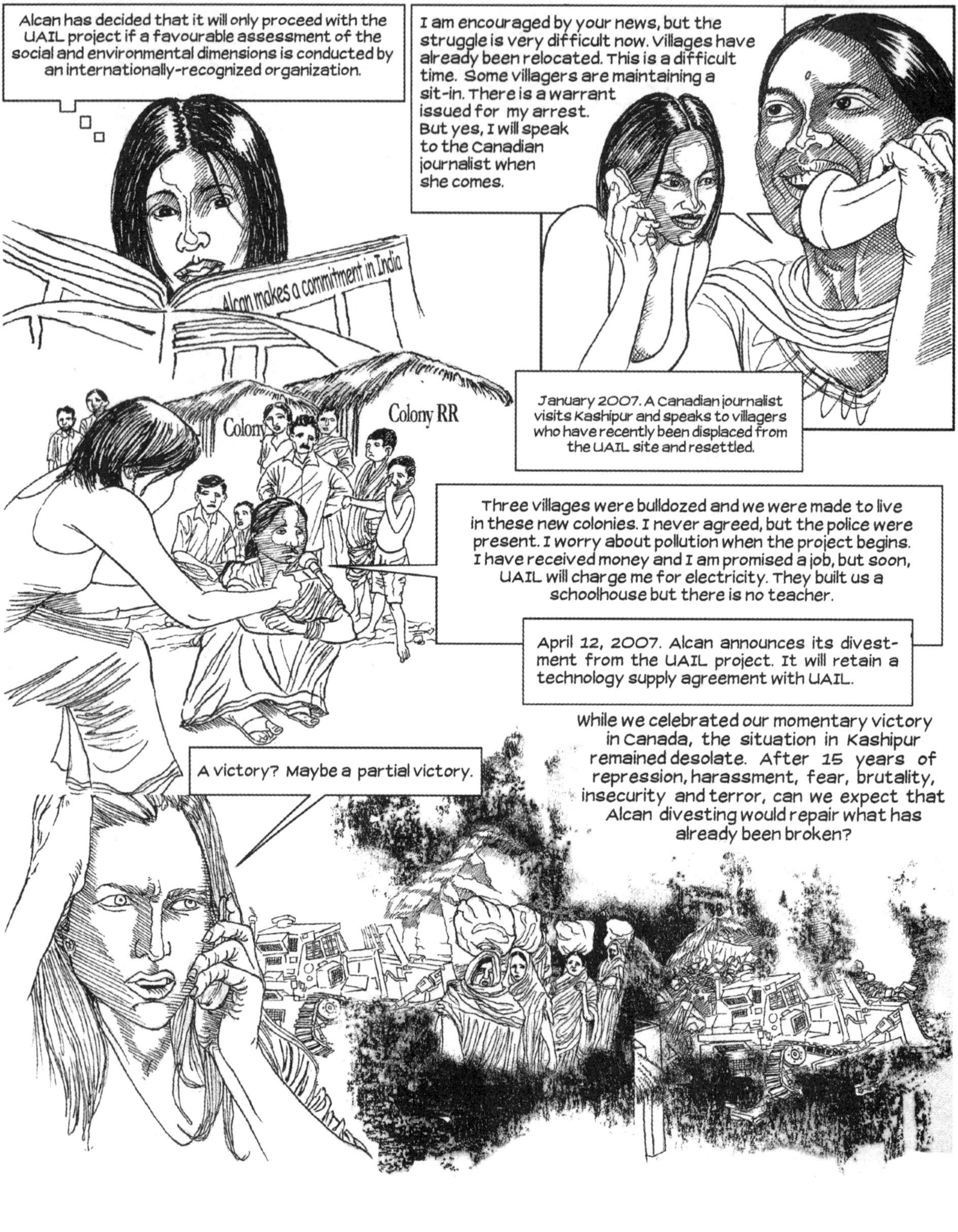
Alcan has decided that it will only proceed with the UAIL project if a favourable assessment of the social and environmental dimensions is conducted by an internationally-recognized organization.
Alcan makes a commitment in India
I am encouraged by your news, but the struggle is very difficult now. Villages have already been relocated. This is a difficult time. Some villagers are maintaining a sit-in. There is a warrant issued for my arrest. But yes, I will speak to the Canadian journalist when she comes.
January 2007. A Canadian journalist visits Kashipur and speaks to villagers who have recently been displaced from the UAIL site and resettled.
Colony
Colony RR
Three villages were bulldozed and we were made to live in these new colonies. I never agreed, but the police were present. I worry about pollution when the project begins. I have received money and I am promised a job, but soon, UAIL will charge me for electricity. They built us a schoolhouse but there is no teacher.
April 12, 2007. Alcan announces its divestment from the UAIL project. It will retain a technology supply agreement with UAIL.
While we celebrated our momentary victory in Canada, the situation in Kashipur remained desolate. After 15 years of repression, harassment, fear, brutality, insecurity and terror, can we expect that Alcan divesting would repair what has already been broken?
A victory? Maybe a partial victory.

The Government of India has a foreign debt to pay. It wants industry. Orissa is resource-rich. Adivasi and Dalit communities are fighting a multitude of extraction projects. They won't even see the aluminum or the steel that is extracted from their lands.
The divestment of one company means the investment of another. Hindalco has already said it is interested in replacing Alcan. Kashipur and the areas around it are parceled off to various companies.
The poorest people continue to fight the most powerful corporations. Without a change in the balance of power, communities living on resources will continue to be "the unluckiest people in the world".
PAURI
GARHWAL
NAINITAL
Mukteshwar
Ramnagar
NAINITAL
Bhimtal
Kathgodam
Jaspur
Kashipur
Haldwani
Bazpur
IN RELATION TO THE REST OF UTTAR PRADESH
In India, Adivasi and Dalits are joined by activists, organizers, lawyers, journalists, academics, researchers, community groups, people's movements, NGOs and others working to change this balance of power and struggling for the self-determination and dignity of the "unluckiest people."
...The Adivasi and Dalit struggle does not stop because Alcan has withdrawn. They have signed an undisclosed "technology supply agreement" with Hindalco, which will likely include refinery and smelting engineering technology.
Alcan Annual General Meeting, April 2007
KASHIPUR NO ALCAN!!

You say that the project didn't meet your environmental or social standards, but you're still selling technology?
We chose to withdraw because the contract outlining rights and obligations for partners does not allow us to influence the course of the project... This is a long term project... We want to keep a long term vision and we don't want to hinder our capacity to return to the market in better conditions later.
What measure does Alcan propose as compensation to villagers for events over the past 15 years of its involvement in India? We are thinking of a compensation package.
The company does not hold any legal obligations to pay damages. According to Alcan, the project was in its feasibility stage.
Who are the unluckiest people?

Kashipur has left a legacy of struggle for communities across India fighting extractive projects. For almost fifteen years, a popular movement led by some of the poorest communities in India managed to stall the operations of one of the world's largest bauxite reserves and one of the industry's largest aluminum producers.
If we continue the solidarity process, it can not only be used for Kashipur tribals but for other tribal communities in struggle all over the world!

OIL

From the Bottom of the Pit

REPORTAGE BY PETR CIZEK
COMIX BY PHIL ANGERS
SCRIPT BY MARC TESSIER & PHIL ANGERS

The oil sands were first mentioned in 1719. The Cree Indians used it to waterproof their canoes.
In 1875, John Macoun led the first government-sponsored expedition of the area along the Athabasca River. He recorded his observation of water naturally washing oil out of the tar sands,
the essence of today's technology for extracting bitumen from the oil sands.
Most of the oil sands lie hidden below northern Alberta's boreal forest, in an area almost the size of Florida.
The first serious effort to dig huge tar pits along the Athabasca river and steam out the oil started in 1963 with the Great Canadian Oil Sands Company developed by Sun Oil Ltd., later to become Sunoco, and eventually Suncor.

By 1967, the company's scion J. Howard Pew, self-proclaimed "champion of free enterprise and enemy of godless communism", had sunk $240 million (over $1 billion today) into this project in an effort to wean North Americans from dependence on foreign oil. However, since separating the oil from the sand and then turning it into bitumen requires huge amounts of energy, steam, and water; this Pew family project was less than successful.
Even after the oil is melted down into bitumen, it still has to be "upgraded" into synthetic crude oil by adding hydrogen usually made from natural gas. Starting in the seventies, the federal and Alberta governments...
...provided billions of dollars in tax breaks and research subsidies, invested in a joint-venture corporation called Syncrude, and even significantly lowered the royalty rate in the oil sands in 1997. In the mid-eighties new...
..."in-situ" technologies such as "Steam-Assisted Gravity Drainage" were developed, which steamed the oil from deposits around Cold Lake and Peace River that were too far below the surface to strip-mine.
By the turn of the millennium, dozens of multi-nationals had invested over $24 billion into the oil sands, which finally began to yield huge profits with the explosive increase in the price of oil.
$24 billion!!! Can you imagine spending that money on developing a clean source of energy instead?

Multi-national corporations, including major players such as Exxon Mobil, ConcoPhillips and Shell also happen to be, by the way, the promoters of the Mackenzie Valley and/or Alaska Arctic natural gas pipelines.
They have confirmed a grand total planned investment of over $100 billion into the oil sands in the next decade. This makes it the largest mega-project complex in the world.
We know there's175 billion "economically recoverable" barrels of oil out of 1.7 trillion total reserves.
Oil production from the oil sands is predicted to quintuple from 1 million barrels per day to 5 million barrels per day between 2003 and 2030, representing over three-quarters of Canada's oil production.
70% of which is destined for export to the United States. In 2001, Dick Cheney's "National Energy Policy" report stated that the continued development of
the oil sands "can be a pillar of sustained North American energy and economic security."

Early in 2006, frenzied speculation hit new heights in a bidding war for new oil sands leases.
Energy corporations spent almost twice as much cash in Feb 2006 than they did for the entire year of 2005.
During the month of January 2006, the Alberta government raised $850 million...
...for selling 4,000 km^2 of new oil sands leases adding this block to the grand total of 24,000 km^2 of boreal forest available for extracting oil from sand.
...A combined area almost as big as Vancouver Island.

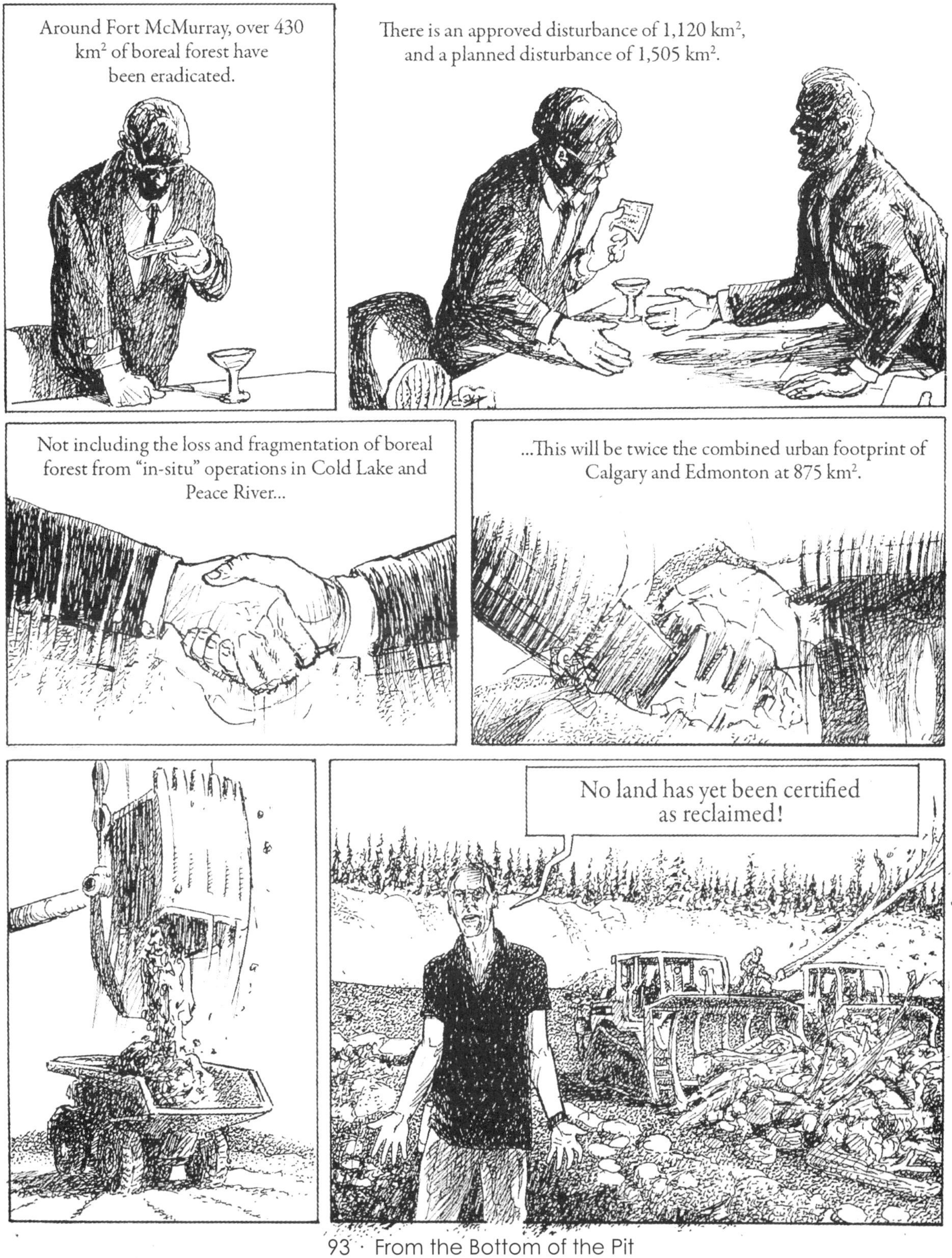
Around Fort McMurray, over 430 km^2 of boreal forest have been eradicated.
There is an approved disturbance of 1,120 km^2, and a planned disturbance of 1,505 km^2.
Not including the loss and fragmentation of boreal forest from "in-situ" operations in Cold Lake and Peace River...
...This will be twice the combined urban footprint of Calgary and Edmonton at 875 km^2.
No land has yet been certified as reclaimed!

To produce one barrel of oil, 4 tonnes of material is mined, 2 to 5 barrels of water are used to extract the oil, and enough gas to heat 1.5 homes for a day is required.
Tar sand producers move enough earth overburden and tar sand every 2 days to fill Toronto's Skydome.

By 2012, they will consume 57 million cubic meter of natural gas per day, enough to heat all Canadian homes for a day.

The greenhouse gas intensity of oil sands production is almost triple that of conventional oil, largely due to the vast amounts of natural gas consumed.
Even before the actual produced oil is burned, carbon emissions from the tar sands are forecast to increase from 23.3 million tonnes per year to between 83 and 175 million tonnes per year.
This represents up to two-thirds of Canada's 2005 "Kyoto Gap" of 270 million tonnes.
Canada's "Kyoto Gap" has increased from 138 million tonnes in 1997 to 270 million tonnes of CO_2 in 2005, largely due to the impact of the Alberta oil sands.

Approved tar sands mining operations are already licensed to divert 349 million m³ of water per year from the Athabasca River.
This is approximately three times the volume of water required to meet the municipal needs of Calgary for a year.
Planned projects will increase water diversions to almost 500 million m³ of water per year.
This represents almost half of the Athabasca River's winter low flow. Can we afford to taint so much water when in the near future...
...water will become the most precious commodity on earth?

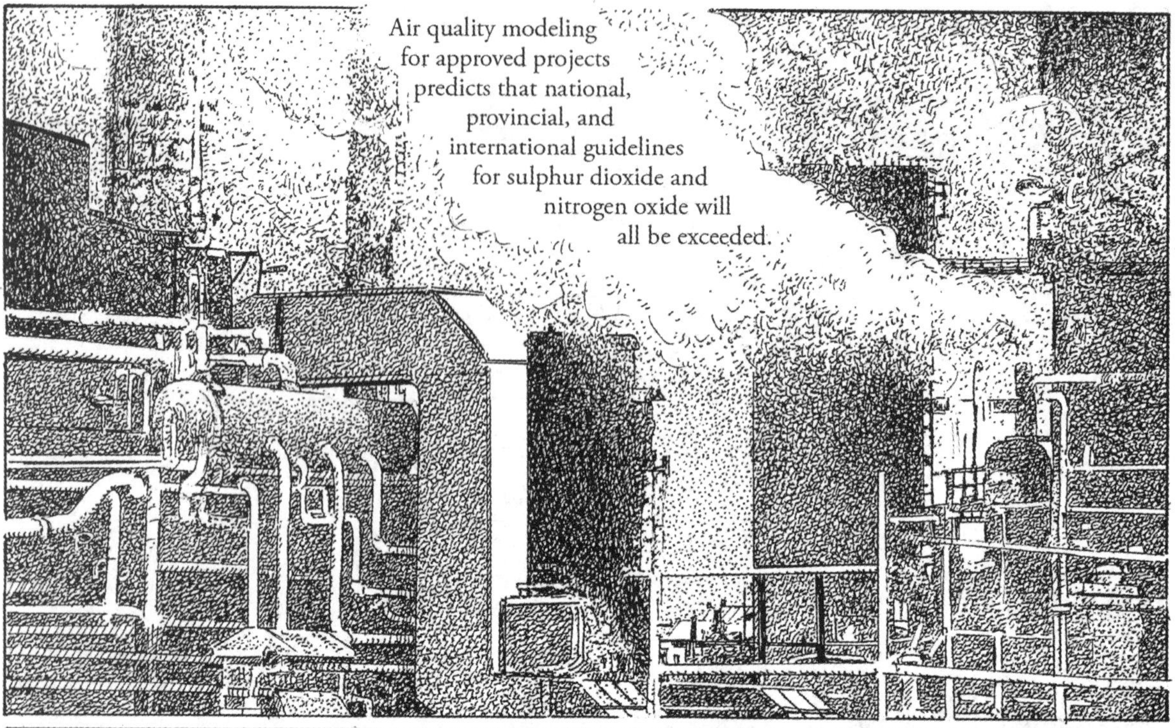
Air quality modeling for approved projects predicts that national, provincial, and international guidelines for sulphur dioxide and nitrogen oxide will all be exceeded.

Syncrude and Suncor are the top two air polluters in Alberta, which have already degraded the once-pristine air quality in Fort McMurray, a small northern city of 71,000, to that of metropolitan centres like Edmonton and Calgary with populations of close to one million each.

The Pembina Institute, a Calgary-based energy think-tank says climate change caused by emissions of green-house gases from human activities is one of the most profound threats to people, economies and ecosystems in the 21st century. Yet, Pembina makes money selling "carbon offsets" to "progressive" corporations.
The money raised subsidizes TransAlta Utilities, a major Alberta coal producer, to construct windmills that produce electricity without emitting carbon. Without legislated "caps" on maximum carbon emissions, the purchase of carbon offsets is a feel-good illusion.
It does not actually result in the reduction of overall carbon emissions. A major purchaser of these credits is Suncor, who recently offset 238,000 tons of carbon out of its total 11 million tons per year of steadily growing carbon emissions.
With Conservative Prime Minister Stephen Harper's announcement that Canada cannot meet its Kyoto commitments, Pembina's climate change specialist Matthew Bramley conceded with regards to a true reduction of carbon emissions:
"We do need to buy international credits as part of our package of measures to meet our Kyoto target."

In November 2005, Pembina produced a report called "Counting Canada's Natural Capital: Assessing the Value of Canada's Boreal Ecosystems".
Its lead author, ecological economist Mark Anielski, proclaimed that the Canadian boreal forest absorbs 173 million tons of carbon worth $1.85 billion each year.
Yet, deep within that same report, Pembina acknowledged recent scientific literature, which concluded that the boreal forest actually became a net source of 44 million tons of...
...carbon emissions per year in the 1970's, largely due to increases in forest fires and pest outbreaks, all related to global warming.

The sponsor of this "natural capital" report is the Canadian Boreal Initiative (CBI), self-described as "an independent organization working with conservationists, First Nations, industry and others to link science, policy and conservation activities in Canada's boreal region."
This "organization" does not have a Board of Directors, is not registered as a non-profit corporation under any federal or provincial laws, and does not have a charitable number.
Yet this "organization" employs a staff of eleven and has sponsored at least 6 major research reports about the boreal forest in the past 3 years.
As noted on CBI's own website, its sole funder is actually the Philadelphia-based Pew Charitable Trusts.
Does that ring any bells?
This same Pew family who originally developed the tar sands, created Suncor, and continues to own Sunoco, a major refiner of synthetic crude oil from the tar sands.

So, how does CBI actually channel money "across the border" if it is neither a legal entity nor a registered charity in Canada?
USA
CANADA
The Pew first transfers money to the North American headquarters of Ducks Unlimited in Nashville Tennessee, which is transferred to its Canadian headquarters in Winnipeg.
Ducks Unlimited, who proudly lists Suncor and Syncrude as its "corporate partners", then recycles "boreal" money not only back to the CBI, but also to other Canadian environmental organizations such as The World Wildlife Fund (WWF) and the Canadian Parks and Wilderness Society (CPAWS).

In Alberta, the once-feisty Edmonton branch of CPAWS is pursuing a similar approach around the tar sands where it is promoting four protected areas, all of which have low oil potential and no leases.

This pattern of environmental organizations adopting a docile "low-hanging fruit" strategy soon after being bankrolled by Pew has been thoroughly documented by U.S. activists and investigative reporters, ...
...Including Jeffrey St-Clair, Alexander Cockburn, Mark Dowie, and Felice Pace.
5M
10$
5M
$

The only bright light is the report "Fueling Fortress America" released in March 2006 by the Parkland Institute, The Canadian Centre for Policy Alternatives, and the Polaris Institute.
This report clearly advocates a moratorium on further tar sands development, a national energy policy, and an exemption —like Mexico—from the "proportional sharing"...
...clause of the North American Free Trade Agreement, which only allows Canada to reduce energy exports to the U.S. in proportion to reductions of our own consumption.
MORATORIUM

Pembina and CPAWS did start advocating a moratorium, but only after Peter Lougheed did so himself. Thus, far Ducks Unlimited and WWF Canada have remained silent on the moratorium issue. Members and individual donors of environmental organizations...
...need to hold these groups' leaders and staff to a much higher standard. Donors need to make sure that they match the rhetoric with the results, instead of just bouncing from one mega-project cash cow to the next, in an endless hustle to line their pockets with lucre.
While countries like Sweden and Iceland are seriously planning to break free from oil by 2020, multi-national corporations are busy digging Canada deeper into a tar-pit, no matter the consequences to the planet's climate.
This is insane, it's self-destructive lunacy!
Yes! And the double tragedy is that so much potential opposition has been readily defused through a wad of cash.
An addiction just as cunning, and powerful as oil.

All this information is in the public domain.
Because he desperately needed a job to feed his family, my brother Joe is working for Shell in Fort McMurray.
This is a complex issue, we are all dependent on oil for goods and services and so many other things.
But if we all voice our concerns.
We will be heard!

Tonight I'm blowing a pile of cash on a piece of tail in town and I'm gonna get so drunk and high man!
Sounds great. You coming with us Joe?
Sorry guys, not tonight. This week's pay goes home to Corner Brook.

MAIN SEC
BUILDI

EPILOGUE

"Our greatest experiment -civilization itself- will succeed only if it can live on nature's terms, not man's."

Ronald Wright

"Life cannot be attained by forcing one's own destiny. One must be receptive to the path laid by nature and circumstance, which will themselves provide what is necessary. The world is ruled by letting things take their course. It cannot be ruled by interfering."

Lao Tzu

The Tao states: "Nature is stabilized by order, and humans, along with all other natural phenomena, exist within nature. Attempting to force one's own path is futile and self-destructive." So I wondered, if one's own path is to resist extraction projects, does protesting "force" one's own path and is it therefore futile. Or is it the natural thing to do to bring balance back?

In the past, I often used my iron will to force things into being. Sometimes it worked but the price I paid seemed, in the end, always steep. Today, I try to curb my desires and be patient with life. But sometimes, one must stand up and intervene.

At the playground the other day, I helped Diego, my three-and-a-half-year-old son, climb to the top of the big slide. As I made my way around to catch him at the bottom, three pre-teen boys with lots of attitude were climbing up the slide, blocking it and preventing the younger kids from sliding down. Before I could even ask them to move away so the children could use the slide, I heard my son boldly yell at them: "Make way, we want to go down!" I smiled. How many times in my life did my sense of justice and bravado get me into trouble! The Tao also says: "Through practicing stillness and receptivity, natural intuition will guide one in knowing when to act and when not to act." This, I strongly believe in and I try to practice.

Top: The three editors of *Extraction!* in the office of Cumulus Press on St-Hubert street, Montréal.
Middle: David Widgington and Frédéric Dubois.
Bottom: Sophie Toupin and Ruth Tait leaving for Mont-Laurier.

When David Widgington approached me with this book project, I had been producing my own books: a peculiar mix of fiction, philosophy and autobiography. Except for taking classes in television production and shooting documentaries while in Cégep, I had not dabbled with journalism since the early eighties.

Extraction! would be a nice break from doing art comix and would provide a stimulating challenge, adapting fact-based journalism into comix. I took up David's offer and joined the project. Art comix gives me the opportunity to flex my voice and reflect upon life as I perceive it. Here my skills as a scriptwriter were needed to help others share their stories through comix.

When I began working for Cumulus, I immersed myself in Dawn Paley's and Tamara Herman's articles, recounting journeys they took to Guatemala and India, respectively, to investigate mining practices. Petr Cizek's feature focused on the tar sands and uncovered financial games on a grand scale. Sophie Toupin's script was an ensemble piece on uranium, featuring multiple characters and a dire warning of things to come in that booming industry.

Top: Phil Angers visits the Canadian Institute of Mining, Metallurgy & Petroleum Show in Montréal.
Middle: Sophie Toupin and Frédéric Dubois on their way to discuss the uranium script.
Bottom: Joe Ollmann and Dawn Paley (left) meet for the first time at the Cumulus Press office, Montréal.

When the first drafts from these four neophyte comix script writers came in, they were filled with statistics, scales, and numbers; calling into question heads of corporations and individuals abdicating their humanity for the sake of maximizing returns on investments. So many facts and so much information! Some had to go. But where to cut? On what to focus? And, these stories were all so damn depressing.

The intention of the project wasn't to turn them into feel good stories (and I doubt we could) but to bring about some balance so that the reader comes away with an understanding of the issues and an awareness of what is ultimately at stake. As a scriptwriter, to marry cinematic images with huge amounts of text and create an interesting visual rhythm introduced some fantastic challenges. To make these stories emotionally resonant, we as editors, moved to emphasize the people on par with the journalistic content. Instead of showing a map, we replaced it with the face of the person providing the information. A human visage has so much more depth, beauty and dignity than any ordinary map. This seemed the key to balance these stories, to keep the focus on real people.

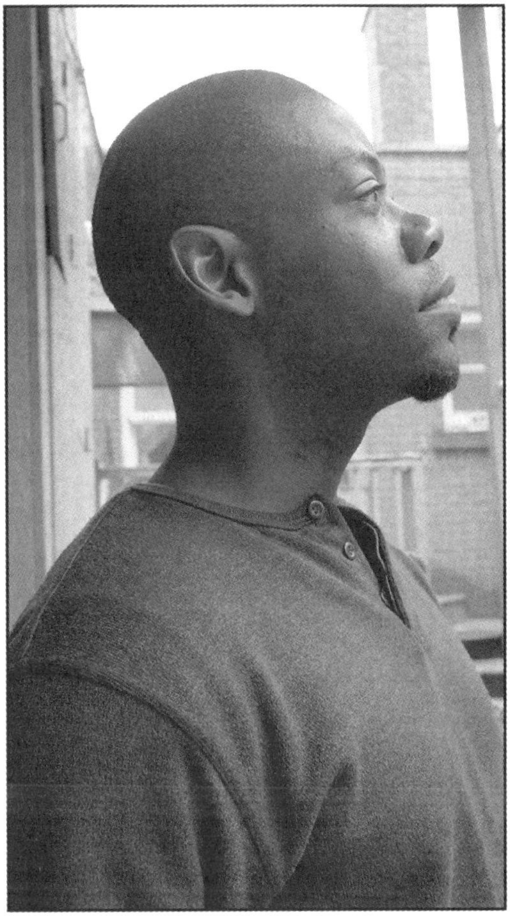

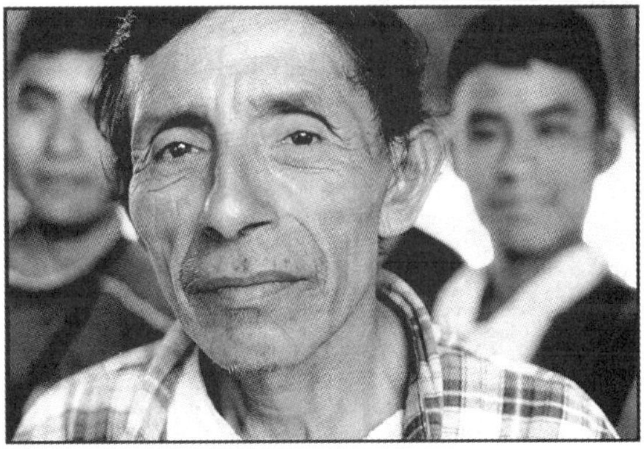

Top: Stanley Wany on the Cumulus Press balcony, Montréal.
Bottom right: An elder from the community of Barrio Revolución near El Estor in Izabal, Guatemala. By Dawn Paley.
Bottom left: A family in Kashipur. By Tamara Herman and Brook Thorndycraft.

"I think one of the reasons that people tend to feel alienated from the political process and from the media nowadays is a surfeit of images that are all equivalent. We have so many mass-produced, commercially oriented images, but very few individually produced images that come from a thinking brain and address a thinking reader."

Françoise Mouly

Cover artist Alain Reno came on board at the start of the project with a rich history of designing award-winning images.

Nowadays, to have an uncluttered cover by an artist without any blurb is a political statement. It is to David and Frédéric's credit that they agreed on an approach to let the artist propose something more personal and striking.

Top: Phil Angers with his first sketches from the tar sands script.
Middle: Alain Reno showing his portfolio to David Widgington before beginning work on the cover.
Bottom: Carlos Santos discussing his concept for the inside cover pages with David Widgington.

As the project moved along, David and Frédéric were given a crash course in comix script writing. They rose to the challenge and made inspired and judicious choices while editing the scripts. As we reached the midway point and the sketched pages came in, we would argue and debate over every comma. As a project with high aspirations and on a shoestring budget, *Extraction!* turned out to be quite a weight to carry. Basically, the salary the artists and the staff received, compared to professional rates, covered the cost of only a couple of comix pages. To ask our contributors to draw twenty pages and make changes every step of the way was a huge deal. The artists turned out to be very generous people and total professionals, giving us their best, above and beyond the call of duty. We were blessed! To spare them more work, we had to work even harder. Yes, we learned how to produce this unique book as we went along. Fusing art with facts and reportage, and turning it into comix, is a magical act. It needs intense concentration to let the brain cells breathe, coalesce and send a spark of creative energy down to the hand that draws. The artists were alchemists, turning ink into gold. Our support and trust in them was essential. It allowed their art to soar. As we moved toward the end, the stories morphed into more than comix reportage: art and words meshed, mashed and embraced. Journalism crossed over into art.

> "In a society divided by class interests, it is impossible to do or say anything that doesn't have political consequences."
>
> *David Fennario*

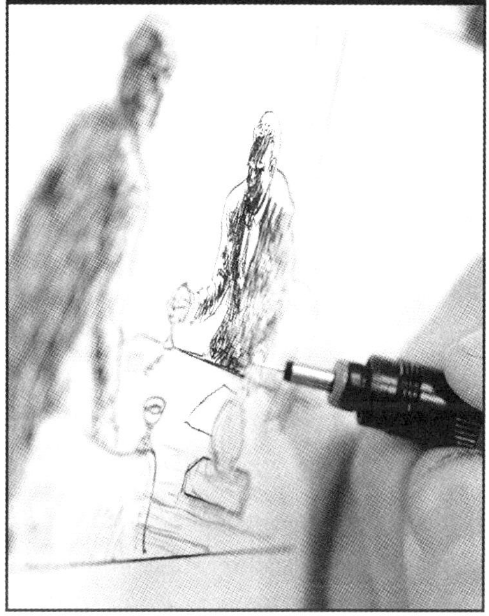

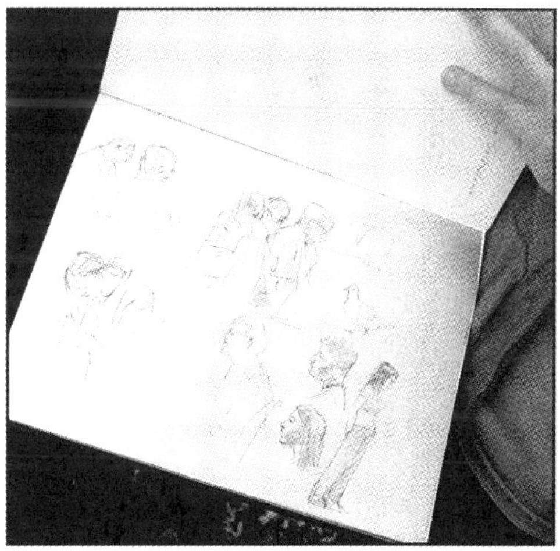

Top: Phil Angers inking the second page of the tar sands script.
Bottom right: A page from Ruth Tait's sketchbook from her Mont-Laurier trip.
Bottom left: Joe Ollmann's final pencil drawings for Dawn Paley's script.

> **"In a time of universal deceit, telling the truth is a revolutionary act."**
> *George Orwell*

"What does it matter how one comes by the truth so long as one pounces upon it and lives by it?", asks Henry Miller. As long as we stuck to the core truth of the piece and were ethical and respectful, we agreed to apply some *mise en scène* to flesh out the characters, which forced us to address the nature of truth. David and Frédéric acted as the guardians of content, striking a rightful balance between my need for stories with real human emotions and their desire for journalistic integrity. In the end everything overlapped. They too were interested in characters as much as I was in the journalistic value.

Frank Lloyd Wright said: "Truth is more important than facts." I could also add that fiction is sometimes more truthful than fact-based journalism. If you want to involve the reader, do not appeal to the intellect, aim for the heart. Art is not an intellectual process, it makes you FEEL! It connects you to the world. It is a tactile, sensual and physical medium. It links you to nature and, says Joseph Campbell, "to the inexhaustible and multifariously wonderful divine existence that is the life in all of us."

Top: Joe Ollmann working on his final pencilled pages.
Middle: Phil Angers inking with reference material.
Bottom: Man on the soapbox played by Luc Gagnon (see final art on page 101).

With our ambitions high and with many input sources, corrections upon corrections came in. Perfection is imperfection. In the end, we had to step back and move on.

A lot of comix, no matter how heartfelt or insightful, fall under the commercial radar and join the ranks of the underground where resistance hides alongside poets, artists and writers. That's not a bad thing in itself but we do hope *Extraction!* gets its chance in the sun since our subject matter could not be more timely. Not many comix reflect upon issues affecting our collective future.

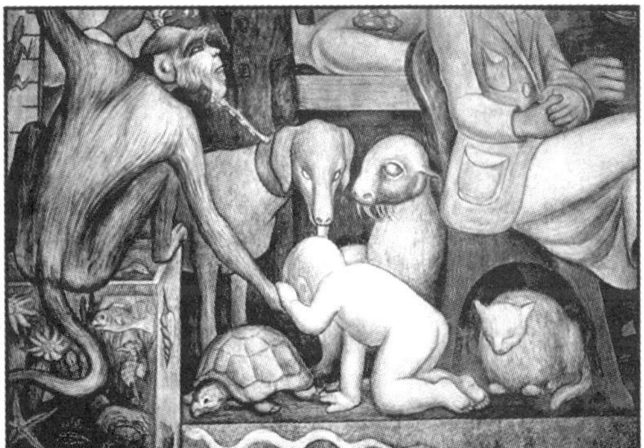

Mexican artist, Diego Rivera, crafted beautiful murals filled with powerful images, inspiring people to band together and bring about positive social change. In 1998, while in Mexico City, I saw his painting, *Man Controller of the Universe.* In a small corner, Rivera drew this new born child protected and sheltered by animals while men and women in fancy suits looked away, enthralled by technological progress. This image has stayed with me to this day.

Marc Tessier, September 2007

Top left: The publisher with Tamara Herman. Top right: Frédéric Dubois and Marc Tessier.
Middle: Stanley Wany and Frédéric Dubois.
Bottom: Detail from Rivera's *Man Controller of the Universe.* A previous version was called *Man at the Crossroad*.

CONTRIBUTORS

Frédéric Dubois is an author and producer of interactive documentaries. He works as the editor of the *Internet Policy Review*. Frédéric has worked on several award-winning productions: *Atterwasch. – A scroll-documentary, Fort McMoney, The Hole Story Interactive*, among others. He worked at the National Film Board of Canada and is a regular production partner of the NFB and ARTE. He completed a Master's in Communication Studies at the Université du Québec à Montréal and a Bachelor's degree in Management at McGill University. As a journalist, he worked for the Association for Progressive Communications and published two books on the topics of media and journalism.

Marc Tessier is a teacher, a comix writer and Photographer. He has edited and designed various anthologies of comix both in French and in English (L'Enfance du Cyclope and Cyclops). He is the author of The Theatre of Cruelty, Mac Tin Tac, Abinagouesh, Petits Nuages de Fumées, À la brunante sur une plage d'agates and Ici sont les Monstres. Together with Stanley Wany, they co-created Éditions TRIP and have been publishing award winning comix since 2009. likeanacidtrip.blogspot.ca.

David Widgington is a cartographer, an archive activist, media artist. He completed his MA Media Studies at Concordia University in 2014. His research-creation project focussed on the role of protest artefacts in developing a movement's oppositional cultural heritage. David co-edited *À Force d'imagination: affiches et artéfacts du mouvement étudiant au Québec 1958-2013 In 2014*, he exhibited, *Printemps CUBEcois*, the first of a trilogy of media installations that explore ritualistic performance theory as it applies to protest. His recent work can be found online at printempserable.net.

Alain Reno was born in Hull, Québec in 1964. Graduating with a BA in graphic design from UQAM, he has worked since 1987 as an illustrator and designer of magazines, books, children literature and posters. He also works for art community, for social organizations and worker's unions of Québec.

Jeff Lemire lives and works in Toronto. He is currently writing and illustrating a trilogy of graphic novels all set in his hometown of Essex County, Ontario. The first two volumes, *Tales From the Farm* and *Ghost Stories* (2007, Top Shelf Productions). The third installment, *The Country Nurse* will be available in 2008. His previous projects include the Xeric award winning graphic novel *Lost Dogs* (2005).

Born in Portugal, it is as a baby that **Carlos Santos** came to live in Montréal. He found his true vocation after discovering the work of comics artists such as Moebius, Franquin and Henriette Valium. Joining the Montréal underground comics scene under the name "d.bilos", his comics have been published in various anthologies and fanzines (*Cyclops, 106U, Stripburger, FishPiss, Image Gun*). He joined in 2010 the EN MASSE crew and his two books (*Raio que te parta!*, 2011 and *Silent Worlds*, 2013) were published by Éditions Trip.

Dawn Paley is a journalist based in Puebla, México. Her first book, *Drug War Capitalism*, was released by AK Press in late 2014.

Cartoonist **Joe Ollmann** is the author of six books: *Chewing on Tinfoil, The Big Book of Wag!, This Will All End in Tears, MID-LIFE, Science Fiction* and *Happy stories About Well-Adjusted People*. He lives in Hamilton, Ontario where art is the new steel, although the steel industry paid better.

Sophie Toupin's work explores the linkages between technology, feminism and activism through ethnographic studies and projects. She presently works for Media@McGill, a hub for research and scholarship on media, technology and culture at McGill University in Montréal. Her current research focuses on focuses on feminist, queer and trans hacker culture, feminist approaches to cryptography and activist algorithms, among others. She is the co-founder of FemHack, a feminist hacklab based in Montréal, whose mission is to create an empowering and inspiring environment for politicized feminist and queer hackers of all gender who want to hack technology, patriarchy, capitalism and other systems of oppression.

Ruth Tait is a visual artist, illustrator and designer working in many different media. She has for many years made the graphic novel an area of exploration and innovation, exploring themes of her own particular areas of interest — relationships, Nature and history. In addition to her freelance work, she is currently working on a graduate interdisciplinary M.A. degree at York University, which engages her in new methods of thinking and experimenting with visual ideas. She is the recipient of two Ontario Arts Council grants for graphic novel development.

Tamara Herman is a community organizer, researcher, filmmaker and independent journalist who has been active in community work related to poverty and resource extraction for fifteen years. Tamara has spent time working with communities living beside mines in India, Mexico, and several African countries. In 2015, she co-directed a documentary called *We Call Them Intruders: A Documentary on Canadian Mining Companies in Africa*. She balances her work as a filmmaker with anti-poverty community organizing in Vancouver, Coast Salish Territories.

Stanley Wany is co-founder and publisher of Les Éditions TRIP. He started his career in 1995 with *Soul Slayer* for the Montréal publisher, Void Comics. He returned to comics in 2006 with a BA degree in visual arts and *bande dessinée* at the Université du Québec en Outaouais where he founded *TRIP* magazine. In 2008, he was a member of the Québec delegation to Angoulème, France and in 2012, he represented Canada at Helsinki's Comics Festival. He has participated in numerous collections including the first edition of Extraction! with Cumulus Press, and the project *Outaouais* for Premières lignes. In May 2015, he published *Agalma*, his first solo book for Éditions TRIP. stanwan.blogspot.com.

Petr Cizek is an award-winning environmental consultant who specializes in the application of GIS mapping and 3D landscape visualisation to land use planning and environmental impact assessment projects. He has over 20 years of practical experience with First Nations and citizens' groups across Canada. Working with the Dehcho First Nations, he helped design and negotiate legal protection for one of the world's largest conservation area networks. Dr. Cizek recently completed a PhD at the Collaborative for Advanced Landscape Planning, University of British Columbia.

Phil Angers (aka Alain Gosselin & Al+Flag) is an illustrator who likes to draw in a realist style with a predilection for politics and the fantastik. With Marc Tessier, Angers created allegorical comics like those in *Mac Tin Tac* or *Krishnamurti's Mountain* (in Cyclope N°1). In *L'enfance du Cyclope*, he presented a chapter from *The Oaken Door*. In this saga, Angers showcases three different visual approaches to tackle three different time periods featuring the same character; the art recalling etchings made on wood or metal. He has always shown an interest in depicting the cracks in the fabric of society. He hopes to finish *The Oaken Door* in the coming years. This work will be his longest graphic novel yet. triskellassocies.blogspot.ca

GLOSSARY

Acid drainage: The outflow of acidic water from (usually) abandoned metal mines or coal mines. Acid rock drainage occurs naturally within some environments as part of the rock weathering process but is exacerbated by large-scale earth disturbances characteristic of mining.

Adivasi: Literally "original inhabitants". Comprise a substantial indigenous minority of the population of India.

AGM: An Annual General Meeting is a yearly conference In which corporations convene stockholders to expose their work done and future plans.

Aluminium: Most abundant of all metals and the third most abundant element in the Earth's crust, after oxygen and silicon. It makes up about 8% by weight of the Earth's solid surface. The chief source of aluminum is bauxite ore.

APEHL: The Association for Environmental Protection in the Upper Laurentians is a non-profit group based in Mont-Laurier, Québec.

Areva: French public multinational industrial conglomerate that deals in energy, especially in nuclear power.

Bauxite: An aluminium ore. It was named after the village of Les Baux-de-Provence in southern France, where it was first discovered in 1821. Bauxites are heated in pressure vessels through which aluminium is dissolved as aluminate.

Biofuel: Broadly defined as solid, liquid, or gas fuel consisting of, or derived from biomass.

Biomass: In the energy production industry, it refers to living and recently dead biological material, which can be used as fuel or for industrial production.

Bitumen: Residual (bottom) fraction obtained by distillation of crude oil.

BJP: The Bharatiya Janata Party, the Indian People's Party, created in 1980, is a major Indian political party. It projects itself as a champion of the socio-religious cultural values of the country's Hindu majority.

Blitzkrieg: An offensive doctrine, which involves an initial bombardment followed by mobile forces attacking with speed and surprise to prevent an enemy from implementing a coherent defence.

Brahmin: Term that denotes both a member of the priestly class in the Hindu varna system, and the name of that caste in the caste system of Hindu society. It is considered to be the highest level in both the varna and caste systems.

Cameco: The world's largest publicly traded uranium company, based in Saskatoon, Canada. It also trades in gold.

Carbon offsetting: Act of mitigating greenhouse gas emissions. The idea of paying for emission-reductions elsewhere instead of reducing by own actions is also known from the closely related concept of emissions trading. In contrast to emissions trading, carbon offsets are voluntary acts without the regulated legal framework of emissions trading.

Caricature: An illustration of a person or character that is exaggerated in its presentation to create a humorous or grotesque representation.

CBI: The Canadian Boreal Initiative is a network of corporate and certain non-profit organisations claiming to protect the boreal forest.

CEO: Chief Executive Officer is the highest and most senior position in a multinational company.

Comix: A movement from the 1960's whereby comic books or comic strips were usually countercultural in nature and made for adult or alternative audiences rather than children.

Composite characters: When a single fictitious character in a story features characteristics of several different characters.

Consulta: Community referendum.

CSR: Corporate Social Responsibility is a concept that encourages organizations to consider the interests of society by taking responsibility for the impact of the organization's activities on communities and the environment in all aspects of its operations.

CPAWS: The Canadian Parks and Wilderness Society is a non-profit environmental group operating "in the spirit of protecting Canada's wilderness".

Dalit: A person often called an untouchable, or an outcaste, who according to traditional Hindu belief does not have any "varnas", which define a person's social status. Dalits fall outside the varnas system and have historically been prevented from doing any but the most menial jobs.

Decoity: Robbery at the hands of armed bandits.

Dene: Group of First Nations living in the Arctic regions of Canada. The Dene speak Northern Athabaskan languages of the Na-Dené language family. They were the first people to settle in what is now the Northwest Territories.

Dividend: A sum of money paid regularly by a corporation to its shareholders from its profits based on the number of shares invested in the corporation.

EIA: An environmental impact assessment is a research document which identifies, predicts and evaluates the biophysical, social, and other relevant effects of development proposals prior to their approval.

Exploration: Act of searching for minerals. It is the first step in the mining chain.

Extraction: Act of removing valuable minerals or other geological materials from the earth, usually (but not always) from an ore body, vein, or (coal) seam.

Footprint: The ecological footprint analysis measures human demand on nature. It compares human consumption of natural resources with the Earth's ecological capacity to regenerate them.

Foreign (direct) investment: Investment made to acquire lasting interest in enterprises operating outside of the economy of the investor. The FDI relationship consists of a parent enterprise and a foreign affiliate, which together form a transnational corporation.

GDP: The Gross Domestic Product is one of the ways for measuring the size of an economy. The GDP of a country is defined as the market value of all final goods and services produced within a country in a given period of time.

Global warming: Concept referring to the increase in the average temperature of the Earth's near-surface air and oceans.

Gold: A highly sought-after precious metal, which, for many centuries, has been used as money, a store of value and in jewellery. Gold is the most malleable and ductile metal and is a good conductor of heat and electricity.

Gram Sabha: Community consultation meeting.

Greenhouse effect: Process in which the emission of infrared radiation by the atmosphere warms a planet's surface.

Greenhouse gas: Greenhouse gases are components of the atmosphere that contribute to the greenhouse effect.

Greenwashing: A term that is used to describe the actions of a company, government, or other organization advertising positive environmental practices while acting in the opposite way.

ILO 169: The International Labour Organization 169 is the convention on Indigenous and Tribal Peoples. It recognizes their right to be consulted before development happens on their territory. Guatemala is one of a handful of countries who have ratified the convention.

INCO: The International Nickel Company, Ltd. is a wholly owned subsidiary of the Brazilian mining company Companhia Vale do Rio Doce (CVRD). Headquartered in Toronto, it is CVRD's nickel mining and metals division. It produces nickel, copper, cobalt, platinum, palladium, rhodium, ruthenium, iridium, gold, and silver.

In-situ technology: Processing method in which the shale is heated underground. This method extracts more oil from a given area of land than conventional oil shale mining and retorting, as the wells can reach much deeper than surface strip-mines.

IFC: International Finance Corporation: Private sector arm of the World Bank, established in 1956. Largest multilateral source of loan and equity financing for private sector projects in the developing world.

Junior company: Small size company most often engaging in exploration work, with little or no assets.

Kyoto Protocol: The Kyoto Protocol to the United Nations Framework Convention on Climate Change assigns mandatory emission limitations for the reduction of greenhouse gas emissions to the signatory nations.

Melton Prior Institute: Germany-based organisation providing the basis for an internationally oriented research on the history of reportage drawing.

Mineral rights: Mineral rights, mining rights, oil rights or drilling rights, are the rights to remove minerals, oil, or sometimes water, that may be contained in and under some land. In jurisdictions that support such rights they may be separate from other rights to the land.

Manhattan Project: Project to develop the first nuclear weapon (atomic bomb) during World War II by the United States, the United Kingdom and Canada.

Natural gas: Gaseous fossil fuel consisting primarily of methane but including significant quantities of other gases. It is found in oil fields, natural gas fields and in coal beds.

Oil: Petroleum or crude oil is a naturally occurring liquid found in formations in the Earth consisting of a complex mixture of hydrocarbons to produce fuel oil and gasoline (petrol), both important "primary energy" sources.

Disclaimer: Most of the definitions below were taken from the open encyclopædia Wikipedia English edition (en.wikipedia.org), double-checked and adapted for this book. This glossary is therefore under the GNU Free Documentation License.

Oligopoly: Market form in which a market or industry is dominated by a small number of sellers (oligopolists).

Open pit mine: Also called opencast mining, it refers to a method of extracting rock or minerals from the Earth by their removal from an open pit or borrow. The term is used to differentiate this form of mining from extractive methods that require tunnelling into the Earth.

NGO: A non-governmental organization is understood as being a non-profit group independent from the government and working on social or environmental issues.

Panzós massacre: On 29 May 1978, the village of Panzós in Guatemala was the site of a massacre in which between 30 and 60 (figures vary) local inhabitants were killed by the national army.

Paramilitary: Group of non-professional armed civilians often affiliated with, but distinct from, a government's armed forces. When they have no affiliation, they are usually referred to as guerillas.

Peak Oil: As first expressed in Hubbert peak theory, Peak Oil is the point or timeframe at which the maximum global petroleum production rate is reached. After this timeframe, the rate of production will by definition enter terminal decline.

PSSP: The Prakrutik Sampad Suraksha Parishad is an Orissa-based (India) social and activist movement opposed to large-scale mining.

Proportional sharing: Clause of the North American Free Trade Agreement, which only allows Canada to reduce energy exports to the U.S. in proportion to reductions of its own consumption.

Radon: Radioactive noble gas that is formed by the decay of radium. Radon is one of the heaviest gases and is considered to be toxic.

Refinery: Industry composed of a group of chemical engineering processes and operations used for refining certain materials or converting raw material into products of value.

Royalties: Common form of payment in agreement to license technology or other assets.

Shareholder proxy: A slop of paper that proves the delegation of power to vote and talk to another member of a voting body. Activists have increasingly used shareholder proxies to get social responsibility, political and faith-based messages out to the assembly of members.

Smelter, smelting: Chemical reduction to produce a metal from its ore.

Spot price: Also referred to as spot rate of a commodity, a security or a currency, it is the price that is quoted for immediate (spot) settlement (payment and delivery). For a security or non-perishable commodity (e.g., gold), the spot price reflects market expectations of future price movements.

Strip mine: Practice of mining a seam of mineral by first removing a long strip of overlying soil and rock. It is most commonly used to mine coal or tar sand.

Syncrude: World's largest producer of synthetic crude oil from oil sands and the largest single source producer in Canada. It is located just outside Fort McMurray in the Athabasca oil sands, and supplies about 13% of Canada's oil requirements.

Tar sands: Common name of what are more properly called bituminous sands, but also commonly referred to as oil sands. They are a mixture of sand or clay, water, and extremely heavy crude oil.

TATA: The Tata Group is India's largest conglomerate, with revenues in 2005-06 of US $21.9 billion, the equivalent of about 2.8% of India's GDP.

Technology supply agreement: In the mining industry, this type of agreement is used by companies wanting to act as technology consultants as opposed to companies engaging in actual extraction. Such an agreement most often goes hand-in-hand with technology supply by the consultant.

UAIL: Utkal Alumina International Limited is a joint venture between bauxite extractive companies operating in India.

Uranium: Metallic chemical element which is approximately 70% more dense than lead and weakly radioactive. It occurs naturally in low concentrations (a few parts per million) in soil, rock and water, and is commercially extracted from uranium-bearing minerals such as uranite.

Uranite: A uranium-rich mineral and uranium's major ore.

WWF: The World Wildlife Fund (also known as the Worldwide Fund for Nature) is an international non-governmental organization "for the conservation, research and restoration of the natural environment".

ACKNOWLEDGEMENTS

Phil Angers thanks Michele for her research and support. Marc for all the beautiful reference photos and for the freedom he was given in adapting the script to comic! A big thanks also goes to the Canadian Institute of Mining, Metallurgy & Petroleum for generously providing reference material for his strip! He he he!

For the insights of uncompromising muckrakers, **Petr Cizek** thanks Felice Pace, Jeffrey St. Clair, Tim Hermach, and Mark Dowie who blazed the first trail over a decade ago documenting the unsavoury links between dirty oil money, the Pew Charitable Trusts, and their Big Green "environmentalist" collaborators in the United States. The great cheer, wit, and humour of my editor Cy Gonick at Canadian Dimension magazine in Winnipeg is thoroughly appreciated. Heartfelt thanks to Linda Henningson at home in Vancouver for all her steadfast love, support, and faith in my search for the truth.

Frédéric Dubois *dankt Rebecca in Liebe für ihn da zu sein*. Frédéric thanks his family and friends in Montréal, Berlin and everywhere else for their support and presence. A resounding thanks goes out to David for his generosity, humanism, guts and devotion. A big thanks to Marc for having driven the book, shared his good spirit, contagious laughter and mystique. Thanks to Dawn, Petr, Sophie and Tamara for the hard work and in-depth research. Thanks to all the comix artists and illustrators for their time and talent. Long live Cumulus Press, independent journalism and comix reportage!

Tamara Herman gives thanks to those in India: Andrew Bachely, Navin Chinchani, Achyut Das, Bidulata Huika, Sibaram Naik, Sudhir Patnaik, Professor Bhagabat Rath, Debaranjan Sarangi, Badal Kumar Tah, and to the villagers of Kucheipadar, Barigan and Maikanch. She also thanks Brook Thorndycraft, Alcan't in India and Barbara Legault.

Joe Ollmann says thanks to Dawn for writing his script and to the people involved in this story for fighting the good fight and setting an example for complacent, cynical saps like himself. To Marc, Fred and Dave for working so hard to get it all together. To *Mad Magazine* and the *gospel of St. Matthew* for forming his adolescent socialist soul. To his family for the love and all that.

Dawn Paley says thank you to people in struggle throughout the Americas, from Six Nations, Grassy Narrows and Sutikalh to Oaxaca, Sipakapa, the Siria Valley and beyond. You give us hope. *A todos los pueblos en la resistencia, desde Six Nations, Grassy Narrows y Sutikalh, hasta Oaxaca, Sipakapa, el Valle de Siria y más allá, gracias. Nos dan esperanza.*

Ruth Tait's thank you's go to Sophie Toupin for being a good companion and intrepid investigator on the trip and to her mother Janice Tait for her encouragement.

Marc Tessier thanks Frédéric and David for the opportunity to work with them on this book. To Marie-Christine for making him laugh in the Cumulus office. He thanks Stanley, Jeff, Alain R., Ruth and Joe for being incredibly generous individuals and a joy to work with. Appreciation goes to Alain Gosselin for being a good friend and agreeing to embark on this crazy project with him. A very special thanks to Heather for supporting his non-profit comix endeavour and to Luc Gagnon for being the best model and friend!

Sophie Toupin would like to thank François Lapierre and all the dedicated members of l'APEHL, Mayor Claude Blain and Mayor Claude Ménard, Gordon Edwards, Véronique Dion, Kasim Tirmizey, Marie Dagenais and Pierre Toupin.

Stanley Wany would like to thank Marc Tessier for giving him this opportunity. Ms Lina Savard without whom he would not be here, Mario Beaulac and Sylvain Lemay for their precious help and, M-C for her support!

David Widgington thanks the following for their support of this book project: Max Douglas, Joe Sacco, David Collier, Chester Rhoder, the Chan Clan, Cédric Piette, Marie-Christine Demers, Abhimanyu Sud, Michael Werbowski, Jeff Lemire, Billy Mavreas, Andy Brown, Chloé Germain-Thérien, Stefan Christoff, Montréal comix jammers, Au Contre-Temps B&B, his entire family, all the MOFs and the POFs, and of course Marc Tessier and Frédéric Dubois for their insight and enthusiasm.

If you are a fan of seeing comics that can talk about social justice, Ad Astra Comix would like to keep in touch! We specialize in comics that promote marginalized voices, and work to dismantle oppressive frameworks. Look for us online!

website	adastracomix.com
facebook	facebook.com/adastracomix
twitter	twitter.com/adastracomics
instagram	instagram.com/adastracomix
tumblr	adastracomix.tumblr.com
pinterest	pinterest.com/adastracomix
e-mail	adastracomix@gmail.com

Thank you to all Indiegogo crowdfunding supporters for helping to make this second edition of EXTRACTION possible!